INFORMATION AND COMMUNICATION TECHNOLOGY

INFORMATION AND COMMUNICATION TECHNOLOGY

Abdul Mannan
Senior Lecturer
Department of Education, Government of U.P.

Himalaya Publishing House
MUMBAI ◻ DELHI ◻ BANGALORE ◻ NAGPUR ◻ HYDERABAD

First Edition : 2007

Published by : **Mrs. Meena Pandey**
for HIMALAYA PUBLISHING HOUSE,
"Ramdoot", Dr. Bhalerao Marg, Girgaon, Mumbai-400 004.
Phones : 23860170/23863863
Fax : 022-23877178
Email : himpub@vsnl.com
Website : www.himpub.com

Branch Offices

Delhi : "Pooja Apartments", 4-B, Murari Lal Street, Ansari Road, Darya Ganj, New Delhi-110 002
Phones : 23270392, 23278631 Reliance : 3018092 to 96
Fax : 011-23256286
Email : hphdel@vsnl.com

Nagpur : Kundanlal Chandak Industrial Estate, Ghat Road, Nagpur-440 018
Phone : 2721216, Telefax : 0712-2721215

Bangalore : No. 16/I (old 12/1), Ist floor, Next to Hotel Highland, Madhava Nagar, Race Course Road, Bangalore-560 001
Phones : 2281541, 2385461 Fax : 080-2286611

Hyderabad : No. 2-2-1 167/2H, 1st Floor, Near Railway Bridge, Tilak Nagar, Main Road, Hyderabad-500 044
Phone : 26501745, Fax : 040-27560041

Chennai : No. 2, Rama Krishna Street, North Usman Road, T-Nagar, Chennai-600 017
Phone : 28144004, 28144005 Mobile : 09380460419

Pune : No. 11, 3rd Floor, Wing-A, Sahadeo Avenue-II, S.No. 5/9+5/10, Someshwarawada, Baner Road, Pune-411 008
Mobile : 9421053747

Lucknow : C-43, Sector C, Ali Gunj, Lucknow - 226 024
Phone : 0522-4047594

Printed at : New Offset Press, New Delhi-110002

Preface

Education is one of the chief concerns of a nation. The higher the ratio of educated citizens, the higher the chances of developing fast. Earlier, the process and the concept of education used to be very complex and cohesive. Various branches of knowledge were to be mastered upon in order to become a successful teacher or a scholar. However, the whole scenario has changed in the present context.

Now, the emphasis is on specialisation. Almost in every walk of life, so many branches and sub-branches of education have evolved and a thorough and comprehensive study of more than one branch would certainly need a prolonged life, which is a thing of the bygone era, now.

Every year, thousands of students are enrolled in colleges and universities. Having attained their targeted academic qualification, most of them seek admission in professional courses in order to get an early job. In a vast country like India, and a mammoth population thereof, it is wellnigh impossible to provide service to everyone.

Of all the job-oriented programmes of study, B.Ed. is a preferred one. It is generally a one-year course, which deals with handling the task of teaching and related matters efficiently. Now, different types of B.Ed. courses have emerged, like simple or ordinary B.Ed. and specialised B.Ed., etc.

Various universities, deemed universities and colleges affiliated to them offer B.Ed. courses. In most cases job is guaranteed, though the exercise, may sometimes be patience-testing. However, sooner or later, B.Ed. degree holders are sure of getting job in one or the other institution.

Apart from the traditional methods of teaching and learning, a B.Ed. degree can, now, also be achieved either on-line or through correspondence. Basically, it encompasses different theories of education, various approaches of teaching, adopted for training students of different abilities, methods of preparing lesson plans, techniques of holding examination and evaluation, etc.

On the whole, it is a rich and interesting subject on which much more has been written so far. With developing technology, new to newer methods and gadgets keep evolving. Here is a series of books for B.Ed. students, as a bold initiative in this direction, which seeks to incorporate the maximum updated information on various aspects of the B.Ed. course in a simple language, and with a sense of authenticity and novelty.

This series consists of several books. This particular book is titled: *Fundamentals of Educational Psychology.*

We hope this meticulous work would prove to be a work of worth. It has everything to satisfy and pacify the urge of the teachers, would be teachers, students, academics, intellectuals, scholars and lay-readers alike.

Yet, we solicit and appreciate your comments and suggestions in order to make it all the more useful and authentic, by all means.

—Editor

CONTENTS

1

Communication and Instruction

In this age of revolution of information technology there has been explosion of knowledge in almost all fields, but man is still discovering, classifying and recording new information at a phenomenal rate starting from the minutest particle of the atom to the farthest body of the universe. Application of this new knowledge in various disciplines has minimized the problems of time and space and facilitated the discovery of still newer knowledge in an endless cycle. Therefore it has been felt imperative on the part of teachers, to keep themselves abreast with the latest developments in their subjects and develop skills in applying criteria to assess the accuracy and validity. The development of new knowledge has resulted in the constant expansion of the school curriculum. As the curriculum expands, not only teachers fail to do justice to teach, but students too get very few opportunities and too little time to learn all objects in depth.

On the other hand, student population is amazingly increasing year by year due to population growth and democratisation of education. Students come from varying experience, backgrounds and differing facilities. Their levels of motivation and aspiration are quite different from one group to another. Even in the developing countries like India a large percentage of students are first generation learners who are deficient in background as well as motivation. Such heterogeneity and diversity among the pupils pose great problems in teaching as well as learning. In view of such problems and developments, special attempts are to be made for two types of instructional media, methods and materials: (i) those which are likely to prove effective for use with groups consisting of individuals of widely varying backgrounds and abilities and (ii) those which can be used advantageously for individualized instruction with only a minimum of teacher assistance.

Fortunately, educational technology has come to our rescue with wide a range of hardware as well as software for tackling these problems. It is the development, application and evaluation of systems, techniques and aids in the field of human learning. The main objective of its use is to optimize learning processes and experience. Technology of education is intended to make learning effective as well as efficient. It facilitates learning through the systematic identification, development, organization and utilization of a full range of learning resources, and through management of these processes. In brief, students are to be enabled to "learn more in less time" meaningfully and to remember longer what they learn without "watering down" or sacrificing quality.

Concept and Basics

According to Wilbur Schramm (1954, p. I) the world famous expert in mass media and communication, there are at least four elements of communication: (1) the source (2) the message (3) the channel and (4) the destination. This applies to both human and mechanical communication. But education is mainly concerned with personal or human communication. In this context either an organization or a person can transmit a message which may be received by an individual or many. The message may be conveyed by verbal expressions, gestures, spoken or written symbols or by hand drawn or photographed pictures. The source is also called an encoder, the message a signal and destination a decoder. These terms indicate the need for common understanding of language and word-meanings based on similar experience background for making communication a success.

Mere presentation of a message or a lesson by the teacher may be a communication, but cannot be effective unless it leads to true learning.

For example, a teacher teaches a new word. The child can repeat the spelling of the word correctly, but he should be able to use the word rightly on subsequent occasions. True learning has not taken place unless the teaching as an act of communication has succeeded in making a permanent and meaningful addition to the student's own communication skills. Brown (1964, p. 9) and others have therefore observed that classroom communication involves far more than the unilateral presentation by the teacher of new words or new facts or new ideas; it requires intercommunication between students and teachers, a mutual sharing of experiences and a continual feedback process.

Norms of Education

Learning means modification of behaviour. Change in response or behaviour is caused partly or wholly by experience. Munn (1954) has rightly defined learning as "the process of being modified, more or less permanently, by what happens in the world around us, by what we do, or by what we observe." Geoch instead of giving a rigid definition of the learning process has made a general statement as following, "Learning, as we measure, is a change in performance which occurs under the conditions of practice." Thus learning depends on the following:

(i) Experience of the learner.

(ii) Capacity and motivation of the learner.

(iii) Methods in which the learning materials are presented to the learner.

(iv) Meaningful or difficulty of learning task; and

(v) Interaction between the above variables.

Sunitee Dutt (1960, pp. 11-12) has enunciated the characteristics of learning as follows:

1. Learning is purposive or goal-oriented. Human beings always learn to achieve something. Purposeless learning may not be effective. But when the goal is fixed and clear, the learning becomes more meaningful and effective.
2. Learning is an active process. There is no learning without self-activity. That is why, it is said, "whether or not you learn and what you learn depends upon what you yourself do, for learning an activity."

3. Learning is individual. Each learner has his own needs and interests, abilities and attitudes, hopes and aspirations. Some may learn quickly and some slowly. Thus the pace and depth of learning depends on that individual.
4. Learning is socially conditioned. It is not only individual but also social. It occurs in relation to social environment of the learner and a teacher has to take this into account.
5. Learning is the response of the whole individual to the total situation. It is unitary as man learns as a whole being. So for effective teaching and learning, the physical, mental, social and emotional aspects of his whole being are to be taken note of and setting up the environment is very important.
6. Learning is creative. It is not just an addition to the previous knowledge and experience of the learner. It is a creative synthesis of all the new and old knowledge and experience. True human learning aims at creative and critical thinking.
7. Learning is transferable. It is transferred to new tasks and the learner establishes relationship of the previous learning experiences with the new ones. Of course it is possible when there is identity of context, procedures and ideals and such transfer is necessary for economy as well as efficiency of learning.

Role of Communication

There are different principles of learning like readiness, exercise, effect and insight which are useful for success of communication. Again there are various approaches to learning. The association theories of learning lay stress on the responses of the learner and bonds with stimuli and the behavioural changes of the learner. The field theories emphasize the learner's perception of the whole field of events. Skinner, Thorndike, Guthrie and Hall are the proponents of association theories. Lewin, Tolman and the gestalt psychologists are the advocates of the field theories. The association theories put emphasis on the stimulus, response, environment and previous experiences and part elements in learning. On the contrary, field theories emphasize on perception and cognition of the learner, the whole rather than the parts and the relationship between parts.

Again, learning processes are classified into three: (i) enactive or learning by doing (ii) iconic learning by organization of perception and (iii) symbolic, learning through symbols and language representations.

All these learning theories, principles and processes are of vital importance for successful communication which is equivalent to active learning. Participation, interaction, cooperation and different skills promote such lively learning experiences. Various media, materials and methods capitalize such principles and approaches for providing better learning experiences. That is why, educational technology consisting of all means, media and materials based on the psychological principles and precepts helps promoting learning as well as teaching.

The Report of the Meeting of Chief Technical Advisers, National Directors and UNESCO specialists in Methods and Techniques, Paris, December, 1979 has said that, Educational Technology is a communication process resulting from the adaptation of the scientific method to the behavioural science of teaching/learning. It is a communication process which may or may not require the use of extending media (*i.e.*, radio, television, film and other audio-visual media). The process contains essentially the following components.

1. Specification of goals and behavioural objectives.
2. Analysis of the characteristics of the learners.
3. Selection and organization of the content or subject matter to be learned.
4. Presentation of the content.
5. Evaluation of the results.
6. Feedback. Feedback must be continuously interacting among all of the preceding five components. It is not just a final component, but it provides essential linkages among all the components.

All these components are also essential elements in learning. Unless all these points are taken into consideration, learning cannot be made effective. These are the basic needs for organization of suitable media and materials in order to optimize learning outcomes. Modern communication methods, visual arts, classroom organization and teaching techniques are to be made use of and geared to achieve the instructional objectives. Communication processes supported by technology of education intend to bring about effective and efficient learning.

Importance of Teacher

Pedagogy, the science of teachers' behaviour, has now given place to mathematics, the science of learners' behaviour, that is there has been a shift in focus from the teacher to the pupil and learners' behaviour has been given more emphasis than teachers' behaviour. In other words, the learning process is replacing the teaching process. Learning to learn is essential and learning is more important than teaching. Learning is also a continuous process. It is lifelong. It is apt to be equivalent with education. Through education students learn how to learn, learn how to keep on learning, learn how to be creative and innovative, and learn how to contribute meaningfully and significantly to solving the new problems that are encountered in day to day world.

If learning is so important, what is the role of a teacher? Faure and others (1982 pp. 77-78) have rightly replied, "the teachers' duty is less and less to inculcate knowledge and more and more to encourage thinking; his formal functions apart, he will have to become more and more an adviser; a partner to talk to, someone who helps seek out conflicting arguments rather than reading out ready-made truths. He will have to devote more time and energy to productive and creative activities such as interaction, discussion, stimulation, understanding and encouragement."

The teacher is no longer the monopolistic purveyor of information and authoritative transmitter of knowledge. He is to inspire, motivate, guide and stimulate. His role is not merely to communicate the new knowledge, understanding and skills but to identify the pupils needs and provide facilities for their all round development.

Elements and Concepts

Communication and its advancement is vital to modern civilization. It has affected the entire world immensely as well as wonderfully. The whole world has been reduced to a close neighbourhood. The distance of time and space has been almost removed and all barriers of castes, communities, creeds and colours have given place to mutual fellow-feeling, interrelationships and cooperative ventures. Robert M-Hutchines, a distinguished educationist has named it "a communication revolution." In fact, today it has exerted inevitable influence on millions and billions of people all over the globe through various means and media. It is informing, entertaining and inspiring individuals in many ways.

Area of Communication

The term "Communication" has been derived from the Latin word "Communis" meaning "Common." It implies common experiences, mutual sharing or give and take. It gives different meanings to different people.

Communication is defined as the sharing of ideas and feelings in a mood of mutuality. It involves interaction which encourages give and take. This provides feedback to individuals involved in the exchange of ideas. Effective communication is, therefore, a two-way process.

Communication assumes a democratic philosophy of life which implies the dignity and respect of all persons who enter into communication. It also assumes freedom to learn and criticize. John Dewey, an exponent of democracy has emphasized shared experience as a key element in communication and pointed out, "communication is a process of sharing experience till it becomes a common possession. It modifies the disposition of both the parties who partake of it."

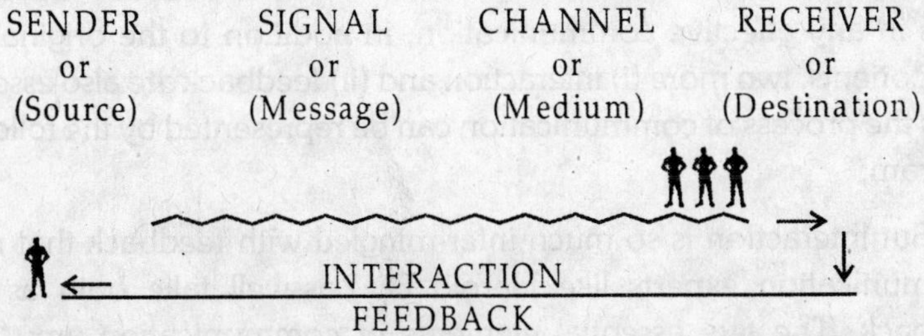

Wilbur Schramm, an outstanding communication expert has said that it "is concerned with all the ways in which information and ideas are exchanged and shared. Thus we are talking about both mass and interpersonal communication. We are talking about the spoken word, signal, gesture, picture, visual display, print, broadcast, film-all the signs and symbols by which humans try to convey meaning and value to another."

Factors of Communication

There are mainly four components of communication. They are: (i) source or sender which is technically called "encoder", (ii) message or signal, (iii) channel or medium, and (iv) destination or receiver which is technically known as "decoder."

In the communication process, the source or the sender has the correct information and transmits the same as accurately, clearly and speedily as possible. The content of transmission is called the message. It should be conveyed through the medium of words both written and verbal, gestures, pictures, and so on. Thus the medium or channel may be either audio or visual or both. Every medium has its peculiar influence on the message to a great extent. That is why, it becomes, in a sense, an integral part of the message and it is aptly called "medium is the message." The sender encodes the ideas or information as the message and channelizes the same up to the receiver through medium or media. The receiver decodes or understands the message or interprets the information according to his background.

Since communication involves at least two or more persons, interaction is a "must" in order to make it effective. This ensures the mutual sharing or give and take of ideas and experiences. Such communication must also have a desired reaction called "feedback". Thus in any effective communication, in addition to the original four components, two more (i) interaction and (ii) feedback are also essential. Thus the process of communication can be represented by the following diagram:

But interaction is so much inter-mingled with feedback that many communication experts like Harold D. Lasswell take both as one-feedback. The five essential elements of communication are: "who, says what, in which channel, to whom, with what effect?"

Important Modes of Communication

There are mainly three modes of communication. They are (i) Speaking ⟷ Listening, (ii) Visualising ⟷ Observing (iii) Writing ⟷ Reading. They may briefly be discussed as follows:

Speaking ⟷ Listening: In this type of communication one speaks and another listens. This is mainly face-to-face as in the case of conversation or listening to a speech. In some cases, there may be gap or distance in between the sender and the receiver or the source and the destination like in radio broadcasting. But both the sender and the receiver of messages share the feelings and may interact with each other if they are in eye to eye contact.

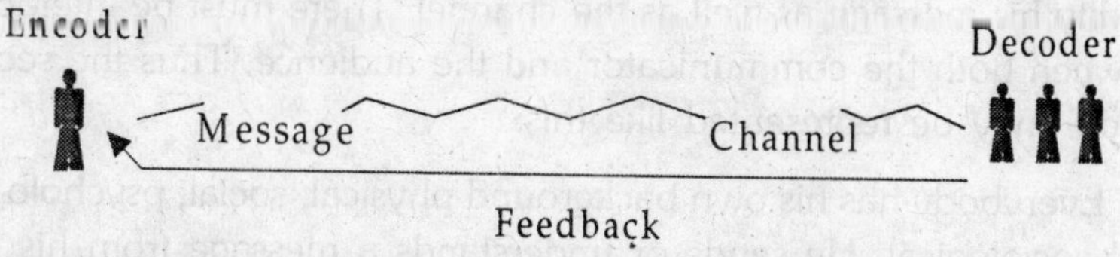

Visualising ⟷ Observing: In this type of communication one visualizes and another observes. There is usually a physical separation between the producer of the message and the audience to whom the message is conveyed as in the case of a television programme or a film. Dramatisation may also be included under this category in which facial expression and gestures produce an effective communication.

Writing ⟷ Reading: In this model of communication, the sender and the receiver are physically separated all the time, but both are able to share the experiences and feelings. An example of this model is the writing of a book by an author and reading of the same by a reader or a group of readers who can enjoy and appreciate the feelings and ideas of the writer properly.

The arrows moving both ways indicate that in all three kinds of communication, interaction and feedback are essential features in effective communication.

Different Concepts

The basic communication theory indicates that there must be four elements—source, message, channel and destination, in the communication process. Any communicator sends message through a channel to the audience. In the technical jargon, the communicator is known as the encoder, the message whether words, pictures or signs is called symbol, the channel may be one of the media or means and the audience is called decoder. The diagram of this basic theory is as simple as this:

In this process, an adept communicator knows what he wants to communicate as message, understands the characteristics of the channel and the social as well as psychological background of individuals constituting the audience.

Then it was found that feedback is essential for making communication effective. The communicator should know the reaction or response of the audience, so that it will be possible on his part to modify his message as well as the channel. There must be interaction between both the communicator and the audience. Thus the second model may be represented like this:

Everybody has his own background physical, social, psychological and sociological. He sends or understands a message from his own background. The physical and emotional conditions like health, mood, attitude and ego involvement influence his understanding or reception of the message. The audience provide their reaction to the message and the communicator modifies or revises his message and media according to such feedback effects. This classic model is again modified by some communication specialists like Wilbur Schramm who advocated a circular diagram to represent the communication process. According to them both the parties—communicator as well as audience—should play the ideal roles as senders and receivers. This diagram is as follows:

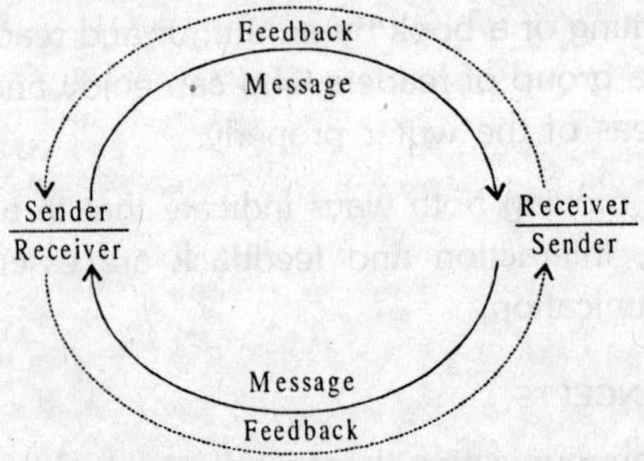

This is a kind of communication that occurs normally in conversation between two persons. But in order to ensure effectiveness in communication, it is felt imperative that there must be interaction and feedback continuously and mutual sharing or give and take of experiences and ideas.

Another theory or model of communication is known as SMCR developed by David K. Berlo. This model emphasizes the psychological nature of communication and provides an analysis of messages and of sensory channels of communication. In this model, source is considered along with communication skills, attitudes, knowledge and socio-cultural context; message includes content, treatment and code; channel consists of eyes, ears, taste, nose and touch and receiver comprises communication skills, attitudes, knowledge and socio-cultural context. The model is represented in the following diagram:

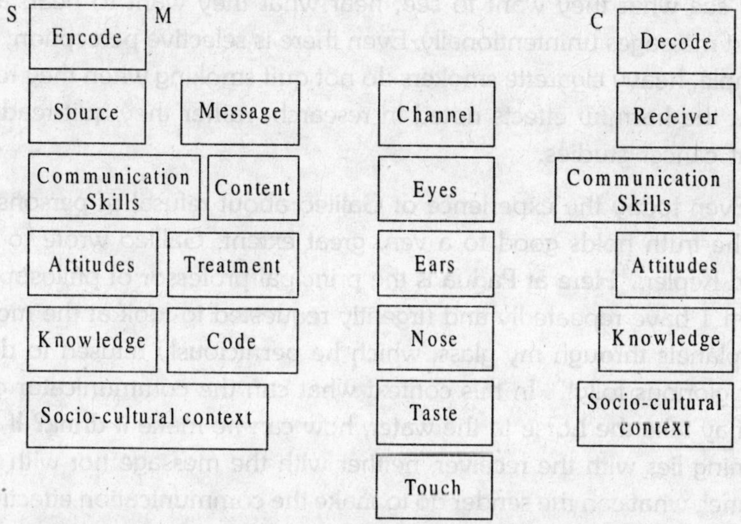

***SMCR Model of Communication by David Berlo*:** Berlo has rightly made it clear that encoding or decoding a message successfully will depend heavily on the communication skills. But the sender's as well as the receiver's attitude is of immenze significance. For example, if the attitude of the recipient is favourable towards the source, reception of the message will be different, if it is otherwise.

Similarly, knowledge and socio-cultural background of the source as well as of the receiver are equally significant for successful transmission and reception of the message. It is a fact that the message actually does not convey the meaning. The meaning is grasped according to the psychological, emotional and sociological background of the receiver. Berlo has aptly said, "Communication does not consist of the transmission of meaning. Meanings are not transmittable, not transferable, only messages are transmittable, and meanings are not in the message, they are in message users."

Significant Queries

It is said that beauty lies in the eyes of receiver. Similarly, meaning lies with the listener. Edgar Dale has corroborated Berlo's statement very strongly and clearly, when he observed that you can present stimuli to an individual, but his needs and background determine whether it will be accurately interpreted, and create the intended meanings. Furthermore, messages are also distorted by the receiver. They see what they want to see, hear what they want to hear, and distort messages unintentionally. Even there is selective perception; for example, heavy cigarette smokers do not quit smoking when they read about the harmful effects noted in research, rather they quit reading about cancer studies.

Even today the experience of Galileo about refusal of persons to see the truth holds good to a very great extent. Galileo wrote to his friend Kepler, "Here at Padua is the principal professor of philosophy, whom 1 have repeatedly and urgently requested to look at the moon and planets through my glass, which he perniciously refused to do... what glorious folly!". In this context what can the communicator do? He may take the horse to the water, how can he make it drink? If the meaning lies with the receiver, neither with the message nor with the channel, what can the sender do to make the communication effective?

Society and Communication

Change in the human society is continuous, perennial and pervading. Change takes place always, everywhere and in all aspects of the society. It may, therefore, be said that the only 'changeless' element in the society is the 'change' itself. That is why *Democritus*, the Greek philosopher has said that you cannot live in the same society a second time.

It is only through change that progress in the society is brought about. More are the changes, greater is the progress. Human civilisation has reached the present stage only due to social changes. The primitive society was comparatively less prone to change than the modern one and the lethargic, dull and backward people are more averse to changes than the energetic, ebullient and advanced persons. Thus the social change is the key to social progress.

Types of Changes in Society

Social changes are mainly of two kinds. One is structural and the other is functional. When the entire structure of the society is changed by means of a political movement or national struggle, it is a structural social change and when a change take place in man's behaviour, ways of living, dress and food habits through education and culture, it is functional.

The American War of Independence, the French Revolution, the Russian Revolution and the Struggle of Indian Independence come under the first category and change over from wearing the untailored dhoti to putting on a pair of suit, from using the bullock-drawn plough to a power driven tractor, from squatting on a mat for eating to dining at a table are the examples of the latter. Both structural and functional changes have almost the same degree of difference as between the major and minor operations. Thus in the struggle parlance 'operation' and in the social science "social change" have both the broad and narrow connotations and have immense significance for physiological and sociological life.

Sociology and Dynamism

According to Ottaway (1962) social change is the result of interaction of both techniques and values. It is of little use to desire social change either in terms of technical progress or in terms of a change in values when one is considered independently of the other. The state of society depends on both and they are inseparable. It is also not easy to say which influence is dominant or takes the lead in any particular social change.

It is said by some social scientists that new inventions and discoveries are a primary cause of social change. But any invention or discovery alone is not adequate to explain the social change. Its use depends on the prevailing set of values in the society. The aims and objectives, lifestyles and philosophy of a community determine how its techniques will be used. The same technical invention can be used differently in different societies or its use may be delayed if the existing conditions of the society are not favourable or it may be accelerated if the society is amply prepared for the same.

On the other hand, some sociologists believe that the value-systems, ideas and aspirations play a dominant part in changing the society. We must, of course, recognize the force of technology, which brings about changes in our habits of work, thought and belief. Thus we have to deal with the problems of social change by considering a number of variables, each influencing the other.

Maciver and Page (1950) in their important work "Society" have distinguished between the technological factors and the cultural factors in bringing about social change. These two factors may be compared with techniques and values given by Ottaway. All these three authors also equally believe in interaction between the two sets of factors and reject all theories which place the technical and cultural factors in opposition to each other and stress the dominance of the one or the other in determining social change.

Ogburn (1947) has also used the terms material culture and non-material culture and described the "cultural lag". When the former comes first, it is more rapid and more effective in determining the social change. That is, non-material factors likes values lag behind the material factors or conditions like techniques. Marx and Engels had earlier put forth their extreme views of economic or material determinism in the change. Recently, Thorstein Veblin has developed a thesis that the kind of work and leisure which men do determine the social structure. Here he has stressed the importance of adjustment to changing techniques.

It may thus be concluded that both material and non-material aspects of culture are simultaneously involved in any social change and that neither is necessarily dominant. The relative significance of these factors will be different on different occasions. The cultural lag or social leeway however often comes into being when a man's values cannot keep pace with his technical development and a cycle of interaction always revolves involving both techniques and values simultaneously, but not equally.

Role of Communication: Communication is the powerful means of bringing about social change. Mac Luhan has developed a thesis of technological determinism which says that changes in the technology of communication brings about social changes. The revolution in the media of communication has helped accelerate the pace of social change during these few decades. Radio, television, newspapers and other mass-media have not only made this world "shrunk" but also have revolutionized the values, attitudes, interests and social milieus.

An important task of communication is to promote development and without its support no project can be made successful. Balcomb (1975) has aptly mentioned that many ambitions projects have failed because of poor all round communications. From his own experience he has clearly pointed out that the UNICEF and other agencies had supplied a considerable amount of audio-visual equipment to various projects but much of this equipment was being poorly used. The hardware was there, but the software had very often been left to take care of itself. That is, proper dovetailing and adequate coordination as well as integration of the technical equipments used for communication purposes and the messages or information that are to be transmitted through these mechanisms are essential for effective communication.

Thus for implementing projects in every field whether health or agriculture or education, the building of such overall communication systems is technically called the "Project Support Communication."

It may be mentioned in this connection that at least four elements are required for communication: (i) a source (ii) message (iii) channel and (iv) destination, (Schramm, 1954).

These involve both human and mechanical segments of communication. There must be person(s) to initiate at one end and receive at the other end and a message to be conveyed by expressions, gestures, spoken or written symbols or by drama or photographed pictures. This process emphasizes the need for common understanding of language and word meaning based on similarity of experience background for effective communication.

The source must have the correct message to be transmitted clearly to the receiver. It is also to be ensured that the receiver is understanding the message accurately and producing a desired response. Thus the communication must be a two way process resulting in intercommunication and based on mutual sharing of experiences and a continual feedback.

Practical Communication

With a view to making communication effective, motivation of the persons concerned is to be generated. The people should think for themselves and on their own initiative. They must do whatever they think right and desirable for them, not as they are told to do. They are to function as project participants, not merely as project's beneficiaries.

Without the needs being felt by the beneficiaries themselves the project howsoever need based and useful, cannot be implemented successfully. A small example may be cited here. In a tribal village women had to go to distant hilltops to fetch drinking water. Under the Community Development Project in that village a well was sunk and sweet water was also available. But alas! The tribal women were so much used to bringing water from a high altitude and not oriented to draw water from a well nearby that the well turned to be a big dustbin!

Another interesting incident of this nature has been narrated by Matheson (1975). Many years back a big project was started in the vicinity of Rome to clear the Pontine Marshes for eliminating mosquitoes and the diseases from the city and at the same time to create more arable land for cultivation of vegetables and fruits. But fishermen living in the swampy places and earning their livelihood from fishing grew irate at the prospect of swamps being drained. As a matter of fact they broke the walls of the water tanks that had been created to lead off the water and soon the ambitious drainage project was completely wrecked.

These phenomena clearly indicate the communication gap as the tribals and fishermen were not explained about these projects which were meant for a better and happier life. Matheson has therefore remarked that unfortunately although the tools of technology are available today to help improve the lives of people throughout the developing world, too many projects are failing to achieve their goals because communication experts and development planners often do not communicate effectively with each other or with the people they are trying to help. Unless the people themselves are made ready for the changes and feel the needs, no project can be a success.

We may use various kinds of mass media and audio-visual aids but our message must be clear at first and communicated as effectively as possible. In order to facilitate this, audience have to be motivated through various means, most important of which is felt to be interpersonal communication. Matheson has again hammered this point when he says that finding the right message for the right medium for a specific audience is a constant challenge to communication experts and development planners and workers. Much has been claimed for the power of the mass media and it has certainly achieved wonders in bringing people instant information and in shaping lifestyles and beliefs. But when it comes to rural societies especially where literacy levels are low and traditions deep-rooted, the effectiveness of mass media in bringing about change is very limited. Experience has shown that probably the single most effective means of communication is the interpersonal one dialogue between two or more people.

Every medium of communication has its own limitations. So it has to be selected according to the socio-cultural conditions of the clientele and be used in appropriate time and place with much caution. Nevertheless communication elements should be built into a project in the initial planning stage. They must be integrated with the whole system and not just tagged on as an after thought. Moreover adequate awareness of the target population should be generated, their needs felt by themselves, their motivations created and attitude changed so that communication is made effective to achieve the desired kind of social change.

Communication by Telephone

In the past, telecommunications has largely been regarded as a natural monopoly in the same way as gas and electricity. The long-term nature of the investment needed to create a national infrastructure, and the massive amounts of capital involved, have discouraged private initiative and relied upon governmental support and backing. During the 1980s, in many countries around the world, this arrangement began to be perceived as too parochial and inflexible; liberalization of tele-communications regulation was seen as a means of stimulating innovation, by encouraging the development of market forces through the introduction of new service suppliers.

Telecommunication and Liberalisation

The liberalization of telecommunication has taken two main forms, they are the following:

— *Privatization:* In which the monopoly authority is established as a private corporation, which competes on the same terms as any other business;

— *Deregulation:* In which the new service operators are allowed to offer competitive services, by the removal of restrictive practices which legislate against this.

Many countries have opted for a mixture of these two forms, often with a transition period setting special conditions which alternative services must adhere to. The extent of deregulation has also differed considerably; since the telecommunications infrastructure is seen as so essential to economic development, the total abolition of all regulatory mechanisms is not considered desirable. For this reason, many people speak of re-regulation, stressing a change in the conditions of government control rather than their complete removal.

The United States has historically been different from most countries, as their telecommunications have always been controlled by a private corporation, licensed under government monopoly. The AT&T corporation which provided the telephone systems prior to liberalization was forbidden in a 1956 ruling from entering the computer or information business, in which it would be considered to have an unfair advantage. In this period, two major corporations, AT&T and IBM, dominated the telecommunications and computing industries respectively, but did not directly compete with each other. A Federal Court ruling of 1982 ended this arrangement by requiring the divestiture of AT&T under the corporate anti-trust laws. This resulted in the creation of seven independent regional Bell operating companies out of the former AT&T conglomerate, each of which provided the public telecommunications service in a separate geographical area of the United States.

AT&T itself remained to compete for the traffic between these regions with other long haul private network operators, such as MCI and US Sprint. AT&T was also permitted to enter the computer and information business in its own right which, as the original developer and licensor of the Unix operating system through Bell Laboratories, was an attractive new direction for its corporate development. This 1982 judgment did not, however, end all regulation over the sector. The divested AT&T was prevented from entering 'electronic publishing'—the provision of information services on their networks—for a period of seven years while competitor network suppliers developed their strengths. This restriction was removed in 1989. The regional Bell Operating Companies (RBOCs) are now permitted to offer:

— electronic messaging;

— audiotex;

— electronic gateways to other information services;

— integrated billing for online information through these gateways;

The AT&T corporation, however, acting as the carrier of inter-exchange traffic, is increasingly interested in developing information services through co-operation with the regional operating companies; for example by using national customer databases of names and addresses for credit card and target marketing schemes.

The exact relationship between monopoly and non-monopoly services has been the subject of debate and conflict between the Federal Court and the Federal Communications Commission throughout this period, with the former championing a more restrictive approach. Realizing that the computer industry which telecommunications serves has global interests, both AT&T and the RBOCs have also diversified into foreign markets, both as telecommunications carriers and as providers of online services, following the creation of new opportunities through liberalization abroad. Although the trend has had an uneven development, there has been a steady blurring of the distinction between information provider, information vendor, and telecommunications carrier which will continue in the present decade.

Europe under Influence

The same global trends have affected the structure and development of the telecommunications and information industries in Europe, but here the influence of the large multinational corporations has been diluted by a more robust government policy concerned to afford some protection to national and regional interests.

Within the European Community (EC) public data networks evolved independently in separate countries, resulting in a smaller overall market for each and a relatively higher cost of equipment than in the United States. This concentration on a national scope has led to the development of somewhat different technology, in particular the installation of different digital switching systems, which reduces the potential for European connectivity as a whole. The European Commission has taken an active interest in the development of the telecommunications facilities within the member countries, where the role of the PTTs has been crucial in the growth from fledgling services to sophisticated international networking.

The EC commissioned the national PTTs to create a packet switching network capable of supporting access from all the member countries, and itself mounted a variety of information services on the network under its own auspices as well as encouraging the development of others. This initiative gave a boost to the developing of information industries in Europe, helping to create a market sector in which information vendors could establish themselves, while the PTTs were engaged in building up their own national packet switching networks to an adequate standard for a reliable international service. Once these had begun operation, Euronet was discontinued in 1984. This period saw the development of Packet Switch Stream in the United Kingdom, and similar ventures in France, the Netherlands, and Germany. These developments were driven through at the national level, with little detailed coordination between network suppliers who increasingly came to view each other as business competitors. This has resulted in the duplication of services and multiple international links to the United States and 'the Far East, with the network suppliers effectively competing to capture the market for data traffic entering and leaving Europe.

The European Commission has been concerned to develop international research and development programmes, and has used the liberalization programme to stimulate innovation and corporate investment, in order to build a compatible European communication infrastructure capable of supporting the open internal market from 1993.

The creation of a European Telecommunications Standards Institute and the progressive reform of the PTT authorities have pursued explicit actions taken to support the *declared EC objectives* of:

— Opening all network services, except the basic phone and data transmission services, to full competition;

— Opening the terminal market to competition, within accepted operating standards;

— Establishing uniform telecommunications charging across the EC;

— Creating European standards for equipment to increase compatibility and remove trade barriers.

To support research and development, the Commission has embarked on programmes intended to develop the economic power of the region through advanced telecommunications technology. The RACE programme is a ten year effort, begun in 1985, to develop the standards and technology for integrated Broadband communication through EC funded project teams drawn from industry and the academic world. The less favoured regions of the EC are supported by the STAR programme, which funds in part the development costs of upgrading communications infrastructures in the more rural areas of Greece, Ireland, Portugal, Spain, France and the United Kingdom.

The development of Narrowband ISDN, however, has not progressed as fast as considered desirable by the EC, which set itself a target of 5 per cent market penetration by 1993 a decade earlier; in early 1992 there were only five commercial ISDN networks in operation in the EC and it was forecast that only 1 per cent would be achieved, failing to impact small and medium sized enterprises and the residential sector. This slow progress is due to a lack of standardization in ISDN technology, a consequent lack of cheap equipment available for the mass market, and an unwillingness on the part of network operators to take committed action in the face of this uncertainty. Within the EC, individual countries have taken a path of deregulation; rather than of complete privatization. Many former publicly owned PTTs have been transformed into private companies, but often with a majority shareholding still held by the State; and although these compete with private corporations in some areas, most still retain a degree of monopolistic privilege and control which in some cases can be quite substantial.

In the United Kingdom, changes in telecommunications occurred within a more general perspective of a Conservative government free-market philosophy committed to reversal of state monopoly industries in general. The successive Telecommunications Acts of 1981 and 1984 separated telecommunications from the postal authority, establishing British Telecom (now BT) as a private corporation.

At the same time, licences to offer similar services were granted to other companies, with the intention of creating a competitive market which would encourage innovation and efficiency. The principal competitor was Mercury Communications, founded for this purpose by Cable and Wireless; two other licences have also been granted for the establishment of cellular radiophone services. A curiosity of the British situation was the third licensed operator, the city of Kingston upon Hull, which for historical reasons already ran its own local phone system; so successfully, that it was able to provide free local phone calls. This was privatized in 1987 as Kingston Communications, eventually becoming part of a larger corporation. The 1984 Act also created OfTel, the Office of Telecommunications, charged with administering the regulatory structure and overseeing the implementation of the licensing conditions imposed on operators, to protect the public interest and ensure fair trading. These included, for instance, a requirement for Mercury to develop a commitment to geographical coverage of the United Kingdom as well as more specialist business services.

Mercury Communications was originally targeted at large corporations, but since obtaining the right to interconnect with the BT network, and thenceforth the ability to provide complete end-to-end services has extended beyond its limited geographic base to provide data services to medium and smaller businesses. As well as digital phone services, Mercury now provides a packet switched data network under the name of Mercury 5000, and an ISDN service called Premier 2100.

In this way, it is now competing with BT in all major areas, and is actively cooperating with foreign companies such as AT&T to provide advanced satellite network services. While Mercury has built an all-digital network from the start but has limited geographic coverage, BT has a completely national service but is only just becoming fully digital in operation. At the same time as upgrading its national network, however, BT has developed international networking capabilities, for instance by the acquisition of the US-based value added network Tymnet, and has developed a range of value added services for delivery on these networks.

There are two levels of integrated digital services, ISDN-2 and ISDN-30, with connections to ISDN networks in many other countries. In conjunction with this, BT provides video-conferencing and 'narrowcast' television services for intra-corporate communications. There is no restriction in the UK on the development of information services by the carrier operators, and BT in particular has developed many such services. The Prestel system was an early venture in videotex or viewdata, but has had little long lasting success despite being one of the first examples in this field.

Although successful in niche areas such as travel and insurance, after ten years of operation Prestel had only 95,000 terminals in use, and had failed to fulfil its original objectives of serving a mainly residential domestic market. Business oriented information services have proved more successful in areas such as financial, currency, market, and company information. Despite deriving some benefit from a loose regulatory structure which in practice has shielded BT from some of the effects of direct competition, the lack of a unifying structure under which services are offered, and a general philosophy of 'let the market decide' has retarded the development of a strong direction for telecommunications in the UK.

The relatively low cost of leased lines, for example, favours the building of private networks and encourages third-party suppliers of value added services, while discouraging the take-up of national services such as ISDN. BT must therefore compete on the same terms as these third party suppliers, and increasingly on an international basis. A further round of liberalization has opened up the possibility of other interested *organizations applying for a licence to offer telecommunications services*. These potentially may include:

— Cable TV franchises, which have extensive links to customer premises, but no national infrastructure,

— Cellular phone services, which are now allowed to build trunk networks, and could offer a packet switched radio-based network;

— Recently de-regulated satellite services, able to both receive and transmit from small and cheap terminal devices;

— Other national industries, such as British Rail, which have extensive rights of way for laying cable.

Broad Connectivity of Telecommunication: The cost of entering the telecommunications market in a competitive age is likely to mean that any of these options would be likely to seek an alliance with a foreign network provider, with particular interest expressed from United States Bell companies.

The relationship between the telecommunications authority and the State has been markedly different in France. The national PTT was established as France Telecom in 1986, a private corporation owned by the State. The organization largely retained control over equipment approvals policy and operating conditions, able, for instance, to levy a 30 per cent surcharge on leased lines used for value added services by competitors and to prevent certain kinds of services altogether. The Transpac X.25 public packet switching network is claimed to be the largest in the world, and has extended its value added services across Europe, including the recent acquisition of the London Regional Transport network in the United Kingdom. Transpac is already used by some United Kingdom companies for facilities such as electronic data interchange, in preference to leased lines from BT or Mercury, and is expected to be a major competitor in the European open market for telecommunications services.

The best known success of the French PTT has been the Minitel service. This videotex service, in contrast to the British Prestel, is now the dominant videotex system in Europe and has been sold to many other network operators, including NTT in Japan, the Italian PTT, and the Bell company US West.

From its first operation in 1983, Minitel grew to 4 million connected terminals in its first five years, and nearly 6 million in 1992, of which at least half are business users. Videotex is generally seen as providing access to large databases of information, presented through an easy to use interface which is accessible without special knowledge or instruction.

The system was originally developed by telecommunications interests as a suitable information service for linking the telephone with a domestic television set, but is now often supplemented by an intelligent terminal or personal computer.

A major reason for the initial success of Minitel lay in the policy of freely loaning the specially designed Minitel terminals to telephone subscribers, at the same time as replacing the printed telephone directory with an electronic version accessible only by videotex.

The Electronic Directory, provides access to a national database of phone subscribers through a simple enquiry language, as well as Yellow Pages information. The great success of this service very quickly created a substantial user base for Minitel, which encouraged other information suppliers to participate by offering *other information services*. These other services now cover a wide range, aimed at both the business and domestic markets, and including:

1. Electronic messaging; professional services; business and financial information; domestic shopping and banking services; and chat lines.
2. The Pancatalogue comprising the online catalogue of the Bibliotheque Nationale and of the SIBIL user group of 22 universities is accessible via Minitel; the new French national library, due to open in 1995, plans a Minitel link which will enable the user to reserve materials from home, and book a reading place for a particular arrival time.
3. Newspaper publishers offer a package of services including news bulletins; information on sport, entertainment and leisure; classified advertisements and games.
4. Professional level services include access to a range of online databases with sophisticated search interfaces. Services are organized in a kiosque system, in which the user has the available services displayed, but pays only for actual use with no fixed subscription fee. There are seven different levels of charging for information used, priced per hour of usage regardless of distance, ranging from almost nominal fees to typical online database

searching rates. The vendor is responsible for mounting the information, and chooses the applicable charging rate. Teletel, the service umbrella under which Minitel is managed, provides access through the Transpac network, and includes the user charge in the normal telephone bill on behalf of the vendor. An international service is being introduced which interlinks with other national videotex services, and with Teletel hosts abroad.

France Telecom opened its Numeris ISDN network in Brittany in 1987, and established a national network by 1990. The development of suitable applications was encouraged by a partnership scheme in which France Telecom sponsored pilot projects with commercial and client organizations, paying half of the research and development costs.

Many of these applications involved the transmission of high definition images, such as the transfer of X-ray and ultra-sound images between medical sites in emergency cases, the transmission of satellite weather images, and the distribution of high definition house pictures around a network-of estate agents.

In Germany the PTT authority was turned into a private corporation, owned by the State, in 1989. Prior to this, the Deutsches Bundespost had resisted the encroachment of competitor services by keeping prices high for leased lines, charging by volume rather than by fixed rental, and by refusing equipment approval.

The 1989 reform following re-unification created a Bundespost subsidiary Telekom, and allowed private companies to compete in establishing value added services, although it has been reluctant to change this policy. The Deutsches Bundespost established a basic rate ISDN in eight major German cities in 1989, and expanded this to a broader national network, using the promotion of applications as a development device in a more limited way than the French experience. *Telekom currently offers the following services:*

— a videotex system, called BTX, similar to Prestel;

— an electronic mail system, called Telebox;

— a facility for remote monitoring of industrial plant and processes.

The development of the telecommunications infrastructure has stimulated the growth of the information industry, creating a larger potential market for information suppliers by bringing many more people into the communications web, building a more reliable and higher quality service.

The European informatics organization, Eusidic, organizes an annual monitoring exercise on the reliability of international packet switching services by asking its members to record a typical week of use. This has recorded a steady increase in reliability, showing a 25 per cent failure rate in calls in 1988 compared with a 14 per cent failure rate in 1991; a large proportion of these faults are believed to occur when a dial-up connection through the PSTN telephone line is used as a local link to the packet switching exchange. An increase in, the total number of packet, switching exchanges in many European countries has reduced the importance of this link, and simplified the installation of new network connections. More dependable communications together with a network infrastructure which is capable of offering a connection to all potential customers has led to a substantial growth in the number of databases produced in both Europe and the United States, and in the number of host systems which offer access to them.

By 1990, there were estimated to be 1048 databases produced in EC countries and 2214 produced in the United States, excluding those only available in videotex model; this represented an 11 per cent growth rate per annum in EC countries and a 7 per cent growth rate in the United States. Within the EC the same report counted 227 hosts offering access to a total of 1256 databases, and 299 online hosts giving access to 3348 databases in the United States, although this is counting some databases more than once when they are accessible from several different hosts.

Compared with earlier surveys, this figure indicates that the number of hosts in Europe is growing while those in the United States are remaining stable. A similar growth has taken place in the professional and business database usage sector of the videotex market in Europe, although the trends here are not so marked.

The French Minitel system dominates the area, with just over 85 per cent of the total number of installed terminals across Europe (6.6 million) in that country; some 50 per cent of the estimated 5 million or so Minitel customers are professional users compared to about 80 per cent of the 100,000 Prestel users in the United Kingdom.

Overall the European Commission estimated that there were 15,000 information providers on European videotex in 1991, although once again most of these are mounted on the Minitel system which provides the predominant means of access to professional information services in France. The distribution of the online market is uneven, with the United Kingdom, France, and Germany accounting for about 70 per cent of the market share, the majority of which is held by the United Kingdom. The large number of potential databases, of which many can often be consulted through competing hosts at varying charging rates and using different command languages as the search interface, is a source of confusion and uncertainty for the end-user.

The diversity of choice has led to a growth in the use of gateways, links through from one system to another. *Gateways in telecommunications networks can take one of three main forms:*

1. Conversion facilities which link dissimilar systems and information types by translating back and forth between the two, typically used for the forwarding of electronic mail between systems.
2. Hosts which have a link through to other relevant hosts in their own area of interest, so that the second can be contacted directly through the first: the user can drop out to the second host, consult the services available, and then return to the first host on completion.
3. Value added services, which provide a 'front end' to a range of databases on different hosts, to ease and enhance user access by masking detailed differences between hosts in a common presentation and interface.

The use of value added networks has been an established part of the online scene for many years, especially attractive in situations where the reliability of the public telecommunications network cannot be guaranteed. The Dialnet value added network used by Dialog Information Services is an example. This offers regular users a local network connection point with billing for telecommunications costs integrated with that for database usage, as an alternative to dealing separately with international telecommunications vendors.

A third-party gateway service takes this a step further by offering integrated access to many different hosts through one connection.

A well-known example is the Easynet service, which allows users to access more than 900 databases on various hosts; the user may select a database directly or enter the relevant criteria for the service to select automatically the most appropriate one itself. The service offers one command language (the Common Command Language) to all the host systems made available, so that the user is not required to learn new search languages and unfamiliar systems are made more accessible. The Common Command Language reduces the degree of sophistication possible in search statements by imposing commonality, so there are facilities for the experienced user to express the search in the native host language. The Easynet gateway also provides management facilities for monitoring the most effective database usage, and for controlling the costs of use throughout the process. This growth in the online information market in both provision and usage is predicted to continue at increasing rates, with market research forecasts continually revised upwards.

The expanding availability of information services and the almost ubiquitous presence of telecommunications networks is beginning to have a substantial impact not only on the way in which traditional information consumers such as libraries and information units are conducting their activities, but also on the development of new information-related occupations which can take advantage of these new communications features.

The practice of tele-working—the conduct of work-related business and interactions over a distance and through the telecommunications networks: with the employer, with colleagues, with customers and contractors—is one of these areas.

The development of information networking, in both the informal and personal sense as well as a formal substitution of electronically-mediated work relations, is fundamental to the growth of distance working. A major plank, in the telecommunications development programme of the European Community has been the encouragement of tele-working through funding of the necessary infrastructure, as a means of economic development in rural and less-favoured regions of Europe. An example of this is the development initiative conducted in conjunction with BT to provide digital exchanges and line plant for the installation of integrated digital services in the Highlands and Islands region of Scotland in the Hi-ISDN project.

A complementary project, called Hi-VANS, provides an easy to use. Interface to national and international services and to the public data network, as well as local information services for the development of tourism and community enterprises. An example of one such enterprise is the use of the network to transmit scanned images of articles which require abstracting from the Elsevier journal publisher in the Netherlands to distance workers in the Highlands. The articles are scanned in the Netherlands from the print version and sent to abstracters in Scotland; the completed abstracts are then returned for publication to the Netherlands with no production delays arising from postal delivery times.

Similar pioneering efforts have seen the foundation of 'telecottages' in many rural areas of the United Kingdom. A 'telecottage' is a generic term for a rural information technology centre, which provides easy access to IT, usually in a public building and on an informal basis. The telecottage can provide formal and informal training, and can act as a catalyst for small enterprises by providing cost-effective access to equipment, as well as supporting the work of community groups.

Apart from public funding, this concept has found sponsorship in the United Kingdom from BT in the form of four Tele-service Centres, established as well-equipped telematic service centres; the first of these to begin activity was in the Shetland Islands in 1991. The willingness of a major telecommunications corporation to become so involved in stimulating the development of a networking infrastructure for information enterprises testifies to the potential possibilities for the future.

In Europe, the concept of tele-working was developed first in Scandinavia, spreading quickly to the United Kingdom, and is now strongly favoured by government and EC initiatives in France, Spain and Portugal. The Telecottage Association, formed to support the growth in tele-working through personal networks of groups and individuals by means of seminars and newsletters, estimates that there are about 590,000 self-employed teleworkers in the United Kingdom. Many of these people are professional workers to whom the production, use and exchange of information is vital to their work.

The recent acceptance of distance working indicates a trend towards more skilled information workers, and the replacement of traditional office work by advisory, and consultancy-based work which does not require full-time attendance at a particular geographical location to be carried out. This is demonstrated by the Tele-medicine project in northern Norway, for example, which provides specialist medical services from a central hospital in Tromsoe to isolated local hospitals and health centres through the telecommunications network. This links together services such as pathology, radiology, and cardiology with access to medical databases and patient information systems for remote diagnosis, without requiring the patient to make a difficult journey to the central hospital facilities.

Generally, *tele-working can take a number of different organizational forms:*

— Home-working, in which the workers are based at their own home, making few visits to an employer premises;

— Tele-centres, which offer new information technology facilities for workers from various different organizations;

— Satellite work centres, which are owned by a large company but are geographically dispersed;

— Flexible working, in which the work is carried out partly at the organizational premises and partly at home.

These new organizational forms imply changes in the overall structures for, organizational communications, and an increased emphasis on the development of networks which are capable of stretching beyond the physical confines of organizational premises alone.

Non-government Network

The backbone of the private network is often provided by leased lines rented on the basis of sole use from the local telecommunications authority. The costs and conditions which are imposed on the lease differ quite widely between countries. Where lines are cheaper, as in the United Kingdom for example, a large corporation may now be able to justify a leased line on the basis of voice transmission only, using one Megastream line to replace several analogue phone lines, and be able to transmit data 'for free' on top of this usage since a fixed annual rent is paid. High bandwidth circuits such as these supply point to point connections, and are appropriate between sites which have a high data traffic flow. Dial-up lines may be used where small amounts of data are sent at infrequent intervals, possibly using parts of a public packet switched network.

Since 1989 organizations operating private networks are allowed to resell spare capacity on their bandwidth, so that linking in with an existing private network already in use elsewhere may be one option; for example, banks may make network capacity available to some of their corporate customers. While terrestrial lines are normally controlled by a telecommunications authority, since the liberalization of services, nonterrestrial transmission offers a wider range of choices. Microwave links can be installed in line of sight between tall buildings, and satellite services can be employed to carry audio, video, and data services within a wide geographical area.

In the United States an 'open skies' policy, under which any qualified organization can apply for a licence to operate a domestic satellite service, has been in operation since 1972; in the United Kingdom, however, the granting of full licences has been more reticent. Satellite capacity is available from a number of specialist data communications companies, and in Europe through the Eutelsat consortium of PTT authorities.

The pricing for Eutelsat usage is differentiated by member country with wide discrepancies in national rates; recently, clients have been able to book capacity through any member offering the best service. The development of very small aperture satellite transmission (VSAT) has lowered the relative costs of this medium, since a receive-only dish uses the same technology as satellite television, and can be installed with the same case without the need for formal planning permission. This can provide a low cost 'point-to-multi-point' data broadcasting, for fast updating of many sites simultaneously; and is also used as a stop-gap measure for establishing a network link quickly, or to link sites which are separated by difficult geographic terrain.

Cellular network radiophones can also be used for non-terrestrial data links, but are not yet ideal for this purpose. This method appears attractive for linking a portable personal computer fitted with integral modem into a corporate network from uncertain locations; however, the cellular channel is noisy, and the transition between cells introduces delays in which contact may easily be lost, preventing truly mobile data communications.

Although the transmission of packet data over the VHF radio waveband is feasible, its widespread use has not yet developed in the United Kingdom; licences have only been granted to radio data companies for quite limited and specialized markets. The wide area networking elements also take on an international dimension if the organization needs to link into sites in other countries.

Larger organizations have diverse information needs, and sometimes specialist but vital performance requirements, demanding the development of a networking strategy which is flexible and controllable. *For information networking to support corporate communications, the organization has three main choices.*

1. To build a private network, using network elements from different sources.
2. To send data over one or other of the public data networks.
3. To hand over all network management aspects to a third party specializing in this field.

These choices have varying significance for different sizes and types of organization operating in different fields; these disparate demands will lead to differing perceptions of the technology which is at the leading edge for a particular situation.

Smaller organizations, or those in which the information and technological infrastructure is less developed, may be well satisfied with options which are already considered obsolete by those in which information has a high strategic or mission importance. The use of public telecommunications in one organization may have as farreaching effects within that context, as are sought by another which is planning frame relay and Broadband networking as replacements for X.25 and ISDN.

Some significant *advantages of the use of public data services* are given below:

1. The development of private networks demand the use of scarce personnel with technical expertize in networking skills, and a high capital investment; while using public networks leaves the organization free to concentrate on its main activities without draining off resources.
2. Once a private network is developed it is hard to change, and expensive to upgrade with technological advances. Information activities may demand a high capacity at critical times, but may not justify this permanently (for example, with seasonal peaks in retail transactions); variable use of the public data network will smooth out peaks and troughs of this nature.
3. The broader scope of public services can develop targeted services for particular needs in a cost effective way, and save on the duplication of effort.
4. In a similar way, organizations can take advantage of the emergence of international standards through participation in a more general service, without the cost of individual development; for example, electronic data interchange standards for basic business documents.

The use of *completely private networks, can have important attractions* especially to a larger organization which is able to amortise costs over a longer period.

1. The strategic importance of information to the organization may demand control over the data at all stages, without fear of compromising security and confidentiality.
2. Control of the network enables technological decisions on speed and efficiency, the use of proprietary protocols, and flexible changes in response to need.
3. Control also extends to economic factors, where the unpredictable nature of changes in public services are avoided while spare capacity can be resold to other clients.

The third choice available to organizations is the use of third-party services to take over the management of all network facilities on its behalf. This removes the need for specialist expertize and keeps the technology up to date, while retaining some control and security, by making this the responsibility of an expert networking company. This is an attractive option to a medium sized organization where communications is important for its business-but not of vital strategic importance—particularly if this involves the use of international connections and the consequent complexity of dealing with different national authorities. Network management companies may offer project management in the development of a network for a specific problem within a defined budget, or full facilities management of all networking arrangements.

Liberalization of telecommunications has encouraged the growth of management services from both private companies and the telecommunications carriers, while the increasing complexity of networking has made their services attractive to the client. The operators have often evolved from organizations which have established their own extensive private networks.

2

Mass Media's Approach

Role of New Technology

Among the most significant forces for change in recent years is the technological sophistication we now possess, for this sophistication not only affects our lives in profound ways but also seems to hold tantalizing promise for increasing our efficiency in education (Kinder, 1973). The last 80 years have seen the development of steam-driven, high-speed rotary presses, advanced optics, films, wire and tape sound recordings, simple and complex duplicating and copy machines, radio, television, computers, and communication lasers.

This technological escalation has bestowed upon education proliferation of equipment and materials which can assist in the reorganization and redefinition of educational experiences. In the past, most teaching depended almost entirely on verbal communication between teacher and student, or written communication to the student from printed materials.

Although, these communication channels continue to play important roles in the learning process, today's students are learning facts, skills and attitudes from pictures, television, recorded words, programmed lessons, and other media. Once technology enters the school building, dramatic renovations usually begin. With the technological magic touch, a simple school-house turns into a systematized learning centre.

Today, many countries around the world use some form of technological media in education. In a few countries, the use is fairly widespread. Most technological devices and programmes, however, are structured around the needs of the teacher and are employed as teaching aids in the classroom. In other words, most educators are using technology to answer the question: how can technology help the teacher? In a few areas, however, focus is on the needs of the student. There, educators ask the question: how can technology help the learner?

In the instances where the student is the centre of attention, technology is catalyst for educational change. Its absence would made a significant difference to the educational process, because technology is an integral part of a well-thoughtout system, not merely a teacher's aid:

Locatis and Athinson (1984) define media as the means (usually audio-visual or electronic) for transmitting or delivering messages. Media includes such things as prints, graphics, photography, audio-communication, television, simulating games and computer.

Schramm, Wilber (1973) in his book *Big Media—Little Media, Aid Studies in Educational Technology,* categories computer, VCR, TV as 'Big Media' and 'radio, filmstrips, graphic, audio cassettes and various visuals' as 'Little Media'.

Meaning and Significance: Nelson Henry has very rightly observed in *Media and Symbols,* "Educational institutions, left to themselves may not be successful in achieving the educational objectives of the developing societies without the support of the new media." New media, implying mass media, as channels of education gain relevance from their capacity to disseminate information to a great number of people (masses) and make the present educational programmes more effective and meaningful.

According to Dr. Marshall McLuhan whose books *The Gutenberg* and *Understanding Mass Media* throw a lot of light on the subject of mass media, it is the medium which is the message. This means that the medium by which a piece of information or knowledge is communicated to us exerts a profound influence on us. The effectiveness of a piece of information depends upon the medium through which it is imparted. Dr. McLuhan thinks that electronic media affect the sensibilities greatly because they tend to massage the sense. Thus, the medium is not only the message but also the massage because it massages the sensory organs and stimulates them to respond actively. Therefore, it is important that the mass media be utilized in the classroom teaching so that the students may obtain sensory stimulation as a part of the process of instruction.

As observed by Wilbur Schramm in his book *'Big Media-Little Media'*, mass media can be made use of in education "as support models in two basic but overlapping ways."

(1) They can beamed part of environment into which learning activities are designed as seen in distance teaching institutions.

(2) They can be brought into the environment as indirect partners or as tools in the hands of the teachers, by supplementing additional or supportive information that is educationally important and useful.

Mass media are means of impersonal communication via some medium, imported through mediated situation. Mass media are means or instruments of communication that reach large number of people or pupils with a common message. The matter may be printed like newspapers or it may take the form of radio, television and cinema. Carlton W.H. Erickson observes, "In recent years technology has swept through society from research laboratories into manufacturing communications, the space age, and finally now, into education."

In early times, the teacher was the only medium of communication for children. He taught his students orally. During the course of time the invention of the printing press, led to the printing of books. Then came newspapers. Now for quite some time new mass media like radio and TV are increasingly used in education. They reach large members and also help in improving the quality of education.

Schools and colleges for long have been the sole medium for imparting information and aiding in the acquisition of knowledge. But with the technological development and fast expanding knowledge, new avenues of education have come up. These media disseminate information which the schools can no longer ignore but needs to be integrated into the teaching-learning process.

The National Policy on Education 1986 and modified policy, 1992 has observed, "The media has profound influence on the minds of children. The mass media make the constraints of time and distance manageable. Modern educational technology must reach out to the most distant areas and the most deprived section of beneficiaries simultaneously with the areas of comparative affluence and ready availability."

Mass media serve some important functions. They are helpful in reaching large number of people. They are helpful in the spread of compulsory education and adult literacy. Recently their use is being made increasingly in distance education. They are useful in making instruction more effective and meaningful.

The use of radio, supplemented by correspondence programme has proved to be very successful in Kenya. Radio forums as a means of adult and continuing education have proved to be extremely successful in Canada, India, Nigeria, Sweden, Tanzania and many other countries.

It is sometimes felt that the mass media tend to diminish the importance of the teacher. It is also claimed in certain circles that they are likely to replace the classroom teachers. A close look at the use of these media of education indicates that they are supplementary media. A lot of work by way of pre-telecast, during telecast and post-telecast remains to be done by the teachers. Many gaps are to be filled up by them. In spite of the explosion of technological media in the developed countries, they still occupy the place of importance. Fears about the replacement of teachers are unfounded.

Radio Technology

There is hardly any doubt regarding the potential of the radio as an instructional aid. Frederic Wittis has rightly remarked: "I like to think of education by radio as a timely, vital and dramatic thing; a system of learning or acquiring more information, a means of widening one's horizon or enriching one's life and breaking down prejudices through inspiration and not perspiration; an education by desire and not by discipline; a pattern of swiftly changing pictures, events with keen interpretations, not statistics and formula's; a moving panorama of the world in which we live—right now, while we are living in it—not a dreary drill of textbooks and tests. In short, I feel that one of broadcasting's most helpful contributions to education and one of its real responsibilities to itself and its listeners is the popularising of education itself."

R.G. Reynolds writes: "Radio is the most significant medium for education in its broadest sense that has been introduced since the turn of the century. As a supplement to classroom teaching its possibilities are almost unlimited. Its teaching possibilities are not confined to the five or six hours of the school day. It is available from early morning till long after midnight. By utilising the rich educational and cultural offerings of the radio, children and adults in communities, however remote, have access to the best of the worlds's stores of knowledge and art. Some day its use as an educational instrument will be as common place as textbooks and blackboards."

Merits

1. Bringing the school into contact with the world around.
2. Helping in the spread of elementary education.
3. Helping in the promotion of adult education.
4. Assisting in the spread of non-formal education.
5. Enrichment of school programme.
6. Furnishing up-to-date material.
7. Developing critical thinking.
8. Developing leisure time interest and appreciation.
9. Providing opportunities for student participation.
10. Providing an alternative approach to the education of out of school children.
11. Imparting vocational skills.
12. Popularising science with a view to developing scientific outlook.
13. Promoting emotional and national integration.
14. Providing information about population education, energy conservation, preservation of wild life, *etc.*
15. Serving as a training component for teachers.

Limitations

1. Radio broadcast is a one-way communication. Students cannot put questions to the broadcaster.
2. The educational value of radio broadcasting depends merely on the use of sense of hearing.
3. The students have little opportunities to participate in the instructional activity. They are passive listeners for most of the time.
4. In several cases, broadcasting time does not suit all educational institutions.
5. The number of receiving sets is not adequate in the case of several educational institutions.
6. It becomes very difficult to integrate school programmes which radio broadcasts.
7. A continuing listening on the part of the students may make them inattentive and uninterested in the task of gaining learning experiences.
8. Usually there is paucity of adequate pre-information, manual or guides regarding radio broadcast with the result that the students and teachers both face difficulties in making necessary preparation for the utilisation of these programmes.

Suggestions for Effective Broadcasting

1. The school broadcasts should not be merely course lessons but should have a wider horizon of application in day-to-day life.
2. The school broadcasts should be planned according to the needs of the syllabus, students and concerned teachers.
3. Teachers should occasionally meet, discuss and plan the type of assistance required on mass media instructional facilities.
4. The radio programmes should be intended to give supplementary information to the various topics in the syllabus.
5. Good planning and administration is highly needed so as to make the programmes effective and worthwhile.
6. Broadcasting time should be suitable to schools.
7. Adequate feedback should be provided.

8. There should be proper follow-up on school broadcasting programmes.
9. Adequate listening facilities should be provided in schools.
10. Broadcasts should be made in easy and simple language.
11. Broadcasts should be made in a pleasing style.

In short all the six main stages of a radio broadcast namely, production, preparation, listening to the programme, feedback, consolidation of acquired knowledge and evaluation should be carefully attended to.

History of School Broadcasting: Ever since the start of school broadcasting by British Broadcasting Corporation (BBC) in 1920, it has made rapid strides in making sound contribution to formal education, In the USA, in 1923, there were programmes in accounting from New York: programmes in arithmetic and literature from Oakland in 1924, 20 States in USA had provision for educational broadcasting. Around the same time about 98% of the schools in UK were equipped with radio and there were regular daily programmes. Bombay Station put out items of special interest to school children. Madras had regular school broadcasts for half an hour on all weekdays. Similar programmes were introduced by Calcutta Station in 1932. The programmes for schools produced by Akashwani Stations are for the following categories of people: (a) Children of primary classes; (b) Children of secondary and higher secondary classes; (c) Preparing lessons for secondary and higher secondary classes near examination times; (d) Teachers; (e) General enrichment programme for children.

These broadcasts can either be 'live' or 'transcribed' depending on the physical presence of the person broadcasting or his recorded speech. Broadcasting organizations throughout the world, including AIR, include in their output various school broadcasts. Such programmes are normally arranged in consultation with the heads of various institutions. This liaison between the radio and educational authorities helps in bringing out effective and useful programmes for the pupils. The planning of such talks is undertaken with great care and by persons of repute. The programmes are prepared termwise and copies are supplied to schools sufficiently in advance to enable the teachers to discuss the subject with the pupils.

TV for Learning

ETV in the World: TV has become child's third parent and a first teacher. The history of television shows that it is a very powerful, informative socializing and mobilizing force. Most of the countries of the world have gone for television to solve their difficulties and problems relating to education. Direct television instruction started four decades back in progressive countries like UK and USA The regular programmes were sent on air in November 1936 by BBC. Remote areas were provided with television sets. By 1958 more than 98% of the population was covered by television transmission. By 1961 Moscow and UK shared the programmes with each other. By 1962 Americam engineers, succeeded in bouncing television waves across the Atlantic on a Satellite-Telestar. In 1967 the first regular service of coloured televisions in Europe began on BBC-2.

For the first time, television for instructional purposes was used in the USA. A large number of experiments in instructional television were conducted there. In 1958, a project entitled continental classroom started instructional television for the whole of USA. It telecasted a programme "Physical of the Atomic Age" for science teachers. About 40,000 teachers received instruction through this programme. Later on, several programmes such as modern chemistry, contemporary mathematics and new biology were also telecast by this project. In 1961, a project called "Mid-west Airborne Instructional Television" started instructional television. About 13,000 schools received the programmes benefiting about five million students, at a cost of 7.5 million dollars or at an expenditure of 1.5 dollar per student. At present there are hundreds of instructional television programmes being telecast in USA and other countries.

Significance of Educational Television: Television is the most potential instrument in educating masses and thereby narrowing down the gap of progress between the developing and developed countries of the world. For a country like India which has vast and inaccessible areas, different climatic conditions, large and ever growing population, TV can be an important central media in providing functional, formal and non-formal education to the masses. It can also help in bringing about social and cultural changes bearing on art, music, drama and literature. It is through television that stimulating and thought-provoking views of renowned statesmen, scientists, educationists, artists and teachers can be shared by all. Television helps in enforcing the public understanding of social, political and scientific advancement of a country. Following are the important merits of television.

Chief Merits

1. It permits the use of the best available teacher to teach a subject for a large number of student viewers. It preserves the expert teaching skills of such teachers on video tape or film for later use.
2. It provides a common experience to all students when they all see the same basic ideas or techniques on television.
3. It provides the teacher an opportunity to observe the instructional methods and ideas of their experts and to increase his own knowledge of teaching methods and stimulate new ideas.
4. It provides technical advantages not readily available in normal classrooms for illustration or demonstration.
   (a) It makes possible close-up magnification of small objects, components, intricate mechanisms, diagrams, *etc.*; giving student a "front-row seat."
   (b) It allows instantaneous change of perspection by switching from a wide camera angle to a close-up or by "zooming" in.
   (c) It permits relationships between two illustrations or time lapse between two stages of a process by dissolving one picture into another.
   (d) It provides for comparison of two or more illustrations by superimposure or 'split screen' effects.
5. It directs the attention of the student to the exact detail of object which he should see by eliminating distracting surroundings.
6. It makes quick and lasting visual and rural impressions which can often reduce the time necessary to teach an idea or technique.
7. It makes it possible to bring large, scarce, new or refined equipment "into the classroom" electronically.
8. It incorporates useful film sequences, slides, graphic art and makes available teaching aids within a television presentation, tailored to meet the needs of a particular course or subject.

9. It saves time, effort and cost of setting up classroom projection equipment.
10. It brings instructional films into classrooms as needed with no special classroom preparation, no darkening of rooms or use of special ventilation in the room.
11. It provides more "immediacy" than instructional films.
12. It brings live demonstration, video-tape or film presentations to the classroom at the instant or immediately after they occur.
13. It permits inclusion of up-to-date information, modification, new equipment or techniques into the classroom instruction.
14. It allows the teacher time to observe individual students or to assist them during the television presentation, or to determine what needs further application after the presentation.

Limitations

1. The medium is limited to one-way communication from teacher to students. Students cannot put any question.
2. The total cost of teaching by television is more than normal classroom instruction, unless television is used to reach large number of students at one time or sequentially over a period of time.
3. Television has special and unique techniques and requires occasional re-arrangement of subject sequence.
4. Individual differences of the students are not attended to in a TV lesson.
5. TV lessons may not suit the school timings.
6. TV lessons are not flexible.
7. Instruction through TV is not child or learner centred.

Kinds of Educational

1. Total TV teaching.
2. TV as a complementary basic resource.
3. TV as a supplementary environment.

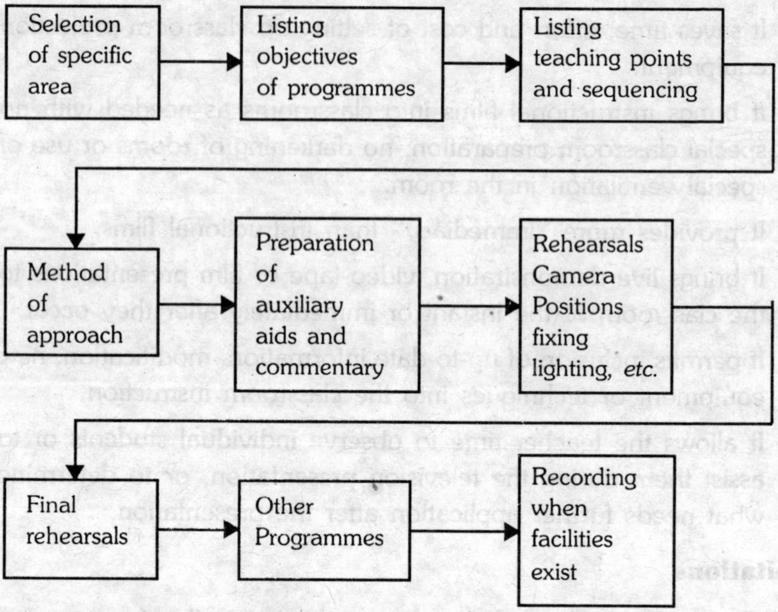

Fig.2.1 **:** Basic procedures in programming ETV.

Kinds of Educational TV

Open Circuit Television: It is the usual type of telecast by commercial or non-commercial stations.

Closed Circuit Television (CCTV): It is the selective telecast which can be used only by specially equipped receivers. Its range is limited to the length of the cable. CCTV can be used to great advantage in educational institutions. Its capabilities are as under:

1. It increases the range of instruction to one or more locations beyond the classroom.
2. It enables institutions to present televised instruction in accordance with their specific needs and schedules.
3. It provides opportunities for the exchange of professors and courses between one institution and another linked to a circuit.
4. In teacher training institutions CCTV with video-tape-recorders can be used to record performance of the teacher trainee during micro-teaching lessons. Video-tape provides the necessary feedback.

5. CCTV is used in many medical colleges in developed countries. The entire operation can be covered by using a single television camera or a battery of cameras located at vantage points.

School TV programmes have five main stages and it is necessary that the teacher should be associated with each stage.

1. Planning and Preparation
2. Presentation
3. Production
4. Utilisation
5. Evaluation

Teacher in the Planning and Preparation of TV Programmes: No TV programme for schools can be planned and prepared unless the classroom teacher is effectively associated with this work. To plan and prepare a good TV lesson, a thorough knowledge of the requirements of the students of different age levels, the suitability of the materials, the sequence and the contents are very vital and this can be met fully only by the classroom teacher. Experience tells that the classroom teacher can contribute effectively in this area if he has a good grounding and knows very well the mechanics of a good TV lesson.

Teacher in the Production of TV Programme: Production is a technical thing but the knowledge about the mechanics of production must be known to the classroom teacher if he is to appreciate a good lesson, *i.e.*, to locate its strong and weak points and suggest improvements.

Teacher in the Presentation of TV Programme: Again, in the presentation of a TV lesson it is only the classroom teacher who can deliver the goods—no doubt a teacher with initiative, imagination and subject competency. The presentation involves only a selected number of teachers but the scope of selection involves all the teachers of a subject. A good selection can be possible only from a TV trained group. It is noted that without necessary training in this field even a very good and competent teacher in a school fumbles and stumbles in the studio.

Utilisation of TV Programme and the Teacher: Pre/post telecast utilisation is the area where the teacher is the master of the situation. It may be emphasized that no TV lesson is complete without the introduction and follow-up exercise in the classroom by the subject teacher. An average lesson with proper and well thought introduction and follow-up can become a very effective lesson whereas a good lesson in the absence of one or both can go flat. The teacher has to inspire the students, prepare them and arouse their curiosity before the telecast of the lesson and afterward has to clarify the doubts, if any, provide the missing links and re-inforcement in the follow-up. He has also to integrate the TV lesson with his classroom teaching. He has also to take care of many other factors and provide congenial conditions to enable his students to really benefit from a TV lesson.

Teacher in the Evaluation of TV Programme: Evaluation is another important area. If tackled properly, it contributes considerably to the quality and usefulness of TV lessons. No evaluation is possible or worthwhile without the involvement of the classroom teacher. Moreover, simply involvement of the classroom teacher does not help much unless he is fully conversant and properly trained to evaluate and assess the TV lesson from all angles and has acquired competency to do the job well. Without proper training the teachers even fail to fill up the check sheets properly—an exercise to be completed by the classroom teacher after every lesson.

Modern Resources

Once the television becomes an integral part of classroom teaching, the physical facilities available in the schools are as important as the quality of the programme, from the utilisation point of view. No matter how rich and valid the TV programmes are, they cannot reach the audience unless optimum physical facilities are available to the viewers. The availability of physical facilities appears to be quite simple yet it requires a constant vigil to keep them all in operational readiness.

Broadly speaking we can classify the physical facilities into three categories:

(a) Space and seating arrangements.

(b) TV sets, Antenna and other accessories.

(c) Literature.

(a) In each TV viewing school, suitable space area must be earmarked for television viewing depending upon the number of viewers. Ideally speaking each class that has to view a TV lesson, should have TV set, fitted in the classroom itself but it is not possible. Therefore provision of a room, which can be converted into a TV room should be made, Hall is another alternative for large number of viewers. Other points to be taken care of are:

(i) Placement of TV set and its adjustment.

(ii) Seating arrangement.

(iii) Lighting.

(iv) Ventilation.

(v) Space for demonstration, per-telecast and post-telecast activities.

(b) It is needless to say that the TV sets should always be in operational condition so that these can be switched on at a very short notice. The various factors that put operating off are:

(i) Defective power points and plugs.

(ii) Defective Antenna.

(iii) Mis-adjustment of TV controls.

(iv) Major defect in the TV.

(v) Voltage fluctuations.

(vi) Operational procedures like locked cupboards, *etc.*

(c) The school television programmes by and large are syllabus based programmes and therefore the students must know the connecting links between their classroom teaching and the television programmes. These should reach the classroom teacher in the beginning of the session or term as the case may be.

The Celluloid

Broadly speaking, a film is a multiple media of communi-cation. It presents facts in a realistic way, dramatizes human relations, arouses emotions and transmits attitudes. It may be used for the communication of ideas, attitudes and experiences to the masses of people. It is very effective in adult education. An educational film has been described as the greatest teacher because it teaches not only through the brain but through the whole body. It has a very powerful influence on the minds of children and is shaping their personality. The main aim of educational film is to elevate and educate them according to the patterns and principles set by the society. A good educational film should help the students to develop a sense of citizenship. A film on national integration can be prepared to help inculcate ideas of oneness and unity. 'Live and let Live' can be that theme. A variety of topics—historical, biographical and of current interest can be covered.

Some of the main advantages of motion pictures are given below:

Increased Reading Interests of the Students: Various investigations conducted in USA show that film producers arouse increased reading interest in children. They are stimulated to get more information about the fact they have observed in a film show.

Real Learning Situations: The film puts before us the learning situations which look to be quite real and actual. The child sees something happening and his experience is direct. Therefore, he is deeply impressed.

Selected Learning Situation: A child learns from his actual life experience more than he learns from the lessons which he gets passively in the school. But life experience comes as a routine.

Sometimes it has no educative value. Sometimes it may have adverse influence. But in an educational film, all the learning situations are selected ones. They have a great educational value.

Quickness of Movement: The events in a film show run very swiftly but without losing continuity and essence of development. Therefore, the effects are also received swiftly in a concentrated form.

Vividness: The learning situation is quite vivid. Everything is happening in such a way as if it is more than living and actual. All the activities are quite vivid.

Motivation: The film motivates teaching. The child takes a great interest. A long study of many weeks may not be able to bring home the acts of the French Revolution to a child but he will at once understand and learn everything about the movement if he is shown a film of the events.

Recreation: A film show is recreational also. It is a play for the child and not work. Thus he learns through playway. He feels light and happy after the show.

Development of Study Habits: Educational films develop study habits as children like to know more about the facts they have leant in the show.

Illustration of all the Learning Situations: We cannot give an adequate idea of an ocean to a child who has never seen a lake even. A child cannot understand what a mountain is like if he has never even come across a hill in his life. But it is not possible for us to take every child to the ocean and mountains, to the deserts and valley, to the Tundras and the Tropics. The film serves us here wonderfully. It brings the ocean, the mountains, the deserts, the valleys the Tundras and Tropics, all in their form and with all their grandeur into the school hall.

Charles, F. Hoban and E.B. Ormer summarise the educational advantages of films as: means of imparting information, development of skills, development of attitudes, enlarging interests and development of the will to solve problems.

1. Educational films sometimes include an element of fiction in historical events.
2. Recapitulation in not possible on the spot. Sometimes the teacher likes and the pupils too wish to dwell longer on a particular shot in the show. But the film does not wait. It goes no.
3. Really good educational films are not available in our country.
4. The whole process is very costly. All the schools cannot afford to have good halls, projectors and other equipment for the purpose.
5. It may have some bad effect on eyesight.
6. Generally teachers are not trained to handle the projector and organize the show.
7. It needs electricity. Many village schools have no electricity.

Suggestion for making the optimum use of Educational Films The most important point to be taken into consideration is that the film must be relevant and purposeful.

When a film show is going to be arranged, the teacher should see that it is really needed in connection with the studies which are going on. He must discuss the background of the learning situations to be presented in the film show, previous to presentation. It should serve the purpose of recapitulation. The whole process may be arranged in the following steps:

1. Preliminary talk about the film.
2. Presentation of the film.
3. Discussion and follow-up on the film.
4. Recapitulation and recording of salient features.

The Children's Film Society India (CFSI) was established in 1955 as an autonomous body with the objectives of promoting and encouraging the Children Film Movement in the country. It also aims at providing children and the young people, films with clean and healthy entertainment. The organization is engaged in production, acquisition, distribution and exhibition of the films suitable for the children and young people. The main office of the Society is at Mumbai and Zonal offices are at New Delhi, Chennai and Kolkata.

Since its inception, the society has produced and purchased about 100 feature films and 104 short films.

A pilot project to reach the rural children in Maharashtra by means of Mobile Film Unit has been in operation for a number of years.

Four children film clubs are functioning, *viz.*, one at Kolhapur (Maharashtra), two at Porbandar (Gujarat) and one at Mumbai.

The society organizes film festivals and participates in important international films festivals abroad. In India, the first international film festival was arranged in Bombay in 1979.

In the year 1981, the Indian festival was accorded 'A' Category status by the International Centre of Films for Children and Young People (ICFCYP), Paris.

The society has set up a National Centre of Films for Children affiliated to the International Centre of Films set up at Brussels under the sponsorship of UNESCO. Some of the important films a teacher can make use are: (1) *Scout Comp*, (2) *Guru Bhakti*, (3) *Ganga Ki Lahren*, (4) *Bachon se Batten*, (5) *Gulab Ka Phool*, (6) *Ekata*, (7) *26 January.*

Newspapers and Magazines

The press is an informal but very influential agency of education. The press includes newspapers and magazines. They provide a variety of information. They cover almost all areas of knowledge. They keep us well-informed. They are very useful in the teaching of various subjects. Instruction through newspapers introduces variety and an element of 'play-way'.

The newspapers are very useful for the study of languages. Pupils learn many new words and many new expressions. They learn how to express themselves and how to follow the expression of others. As regards social studies they learn how the society is developing day by day. They learn a lot about the society. There is much geographical and scientific information also in daily papers and magazines.

For international understanding the study of newspapers is essential. Children come to know how the world is progressing, how we are woven internationally, how the events occurring in one country affect all the other countries of the world and how we shall have to suffer if the Third World War breaks out.

In the teaching of arithmetic, the newspapers can furnish examples concerning banks, interest on savings accounts, deposits is also very helpful in the teaching of economics and commerce. Likewise a lot of information on various subjects is available form newspapers which can be used in daily teaching.

3

SIGNIFICANCE OF INFORMATION

Information is the life-blood of a community information service and the information file is at its heart. Unless the heart is sound and continually pumping a supply of regularly renewed and fresh information into the system, it will not function at its best. Therefore, it is important to give extra care and attention to planning the resources needed to set up a sound information base and a workable system for keeping it up to date.

Some basic things you will need to consider are as follows:

— How is the information to be collected?

— What method is to be used for processing and storing information?

— How is the information to be retrieved?

— What size is the file likely to be?

— What areas, geographical and subject, is the file to cover?

In addition, you will need to consider how the information is going to be disseminated, if it is not just for internal use. The kind of system you choose and the degree of its sophistication will depend on the nature of your information service, the size and complexity of the information you need to store, the staff, and the financial resources available. When designing a system, make sure it is readily understandable to all those using it-volunteers and public included-and not just to professionals.

Wide Scope

In identifying the need for your community information service, you should have reached a decision already on the community to be served. This may be either a geographical community or a community of interest or both. Either way, you will still need to make a decision about the extent of your information base. For example, a neighbourhood information service must decide what information to collect on the wider area-town, county, region, nation-outside its local community. An information service aimed at a fairly wide area may need to limit its interests to a particular sector of the community or to a particular range of subjects, *e.g.*, the low paid, those receiving income support, or those with debt problems. Even an information service in support of a local campaign may benefit from collecting information on similar campaigns or problems country-wide.

Kinds of Information

The information base of your service is most likely to include the following types of information:

Soft information—details of clubs, societies, organizations and services; individuals; events, *etc.* Usually this information will not be available in a published form, or at least not in sufficient detail, and it will be necessary to make a conscious effort to collect it. This information may comprise the major part of the resources file.

Hard information—factual information on a specific subject, *e.g.*, benefit rates for single-parent families, how to get legal aid, how to change your name. This information will be available in a variety of Banns, some ephemeral such as leaflets, pamphlets, booklets, broadsheets, posters, periodical articles, *etc.* Such items are usually kept in a vertical file, in storage boxes or similar receptacles and accessed via some form of index or classification scheme. Other hard information may be in the form of books, multi-volumed reference works, law books, or even audiovisual forms. If there are many items of this type, they will need to be classified and shelved.

Supplementary information—information already produced in a particular format-directories, handbooks, diaries, annual reports, constitutions, newsletters-of organizations appearing in your information file.

Importance of "Soft" Information

It is highly unlikely and undesirable that an information service should need to 'go it alone' in the collection of information. Such is the volume, complexity and variety of sources of information in present-day society that, unless your service has very narrow terms of reference, it will be virtually impossible for it to collect and keep up to date all the information that is required.

Therefore, it is important that you first of all identify the information providers, support services and 'gatekeepers' in your community and establish effective contacts with them. These links may already have been forged in the process of conducting a community profile. They now need to be fostered and strengthened, so that there can be a mutual exchange of information. These contacts will also be able to provide you with useful feedback from the community as to the success or otherwise of your service. The network of contacts can be maintained on a fairly casual basis as the need arises, or you might want to formalize the arrangement. Some ways of doing this include irregular 'get togethers', informal luncheon clubs, regular meetings with agendas and minutes, or the circulation of a newsletter or bulletin. Other activities that could develop out of such meetings are joint collection of information, shared publicity, compilation and publication of directories, information handbooks and leaflets, training, and general discussion of common problems. Collecting information is a time-consuming process and there is no one method of going about it. You will probably have to use a combination of several techniques to build up a satisfactory information base. One thing definitely to be avoided is duplicating work that has already been done. So first of all identify the following:

Existing Information Files: Contact all the other community information services, council departments and organizations who are likely to maintain information files and ask if they would be willing to share this information. Try to offer something in return, either an exchange of information or some other help you can provide.

Local Directories: There are several types of directories that may be available in your area and which are useful sources of soft information: (a) local government authorities often produce directories or town guides of their area which contain a certain amount of local information or they may produce a guide to council services; (b) telephone directories, especially Yellow pages which have an alphabetical subject arrangement; (c) area directories, such as those produced by Thompson's Newspapers, which are similar to Yellow pages but also contain a section of community information-your local newspaper may produce a directory of local services, sometimes in conjunction with local information groups; (d) citizen's guides may also be produced by the local newspaper as a supplement to an existing newspaper; (e) CVSs or RCCs sometimes produce directories of voluntary organizations in their area or you may find these listed in your local CVS/RCC annual report; (f) directories covering a special subject, such as accommodation, halls for hire, *etc.*; (g) directories aimed at a particular interest group or groups, *e.g.*, people with disabilities or senior citizens.

National directories are useful not only for the wider network of services but also for identifying local offices of a national organization.

Other methods to use in building up your information base and maintaining its currency include: Looking and listening. A lot of information can be gathered by simply walking around your community, looking at notice-boards, picking up leaflets, or attending open meetings, fairs, fetes and other community events. This method of collecting information is unpredictable and requires more than a 9-to-5, five-day week.

Contact with individuals. People are a major source of information in any community and there are always those to whom others naturally go for information. Try to identify such people and enlist their help, possibly by offering something in return-bartering is a time-honoured system! Direct contact with individuals in other information services and organizations is often a better way of eliciting information than impersonal methods and also paves the way for further cooperation. Pay regular visits to such centres.

Scanning newspapers, magazines, newsletters, *etc.*, can turn up details of new organizations; changes of personnel, premises or hours of opening; new services introduced; and may highlight problem areas where information may be needed.

Other documents, including council minutes and agendas, planning regulations, annual reports (ask to be put on mailing lists for these), leaflets and manifestos.

Media publicity Local press, radio and cable television stations may be prepared without charge to put out a call for organizations to contact your service. If not, then consider inserting an advert or, where there is, a local directory produced, have a tear-out slip included for organizations to fill in and return to you. As a result of these various methods, an amount of raw material will have been obtained which may be of use to your information service. How accurate, up to date and complete the data is will depend very much on the reliability of the source.

In most cases, unless you have every confidence in the source, the information will have to be checked before it is entered in a more permanent form in your resource file. So, at this stage, simply record it onto say a scrap 5in x 3in card and put it into a temporary file.

The information you have obtained will be of several different kinds but the most common will probably be that relating to organizations, clubs and societies. This is the most difficult type of information to collect since it is ever changing and therefore unrelenting effort will be required to achieve even reasonable completeness or accuracy-100 per cent will be impossible. However, with a good system for collecting and updating the information, you should be able to build up and maintain an acceptable file.

A lot of staff time will be needed to carry out this work, so first of all investigate whether there are other local information services that need this type of information and would be willing to help in its collection. It may be possible to collect the information through an umbrella group, where the work can be framed out so that it is not too great a burden on any one organization. Alternatively, a joint approach can be made on behalf of a number of organizations.

Ideally, the most effective way to collect information of this kind is to pay a personal visit to each organization or service, but rarely will that be possible without extensive staff resources or a very small community. The cheapest and quickest way to collect this information is by telephoning. Prepare a standard list of questions to ask so that entries in your information file follow the same format.

Telephoning is only practicable if you have a reasonably small number of organizations to contact, say up to one hundred. If visits and telephoning are ruled out, then the information can be collected initially by postal questionnaire, with a request for up-to-date information sent at least once a year and also when it is known from other sources that changes have taken place. A covering letter should be sent with the questionnaire or update request, explaining "Why the information is needed and asking for cooperation. If the information is to be passed on to other organizations or commercial bodies, either free or for payment, this should be drawn to the attention of the recipient of the questionnaire, so that they have an opportunity to decline to give the information or insist that it is not so disseminated. If the information is to be stored in a computer, then you will need to meet the requirements of the Data Protection Act.

If it can be afforded, include with your questionnaire some means of returning it free of charge, as this will significantly improve the response rate. Separate forms may be required to collect different kinds of information. The following is a suggested list of headings which you might need to include on forms for collecting information from two of the most common types of organizations.

Clubs and Societies

- — Name: the name by which the club is best known, plus the full name if different. Indicate relationship to larger body, *i.e.*, branch, regiment, lodge, *etc.*
- — Secretary's name, address and telephone number or those of a similar person if the organization has no secretary.
- — Place, day and time of meetings.
- — Purpose of organization if not self-evident from the name.
- — Eligibility: any restrictions by age, sex, ethnic group, occupation, status (single, divorced, widowed), *etc.*
- — Subscription: any charge to become a member or to attend meetings.

— Annual general meeting: the date of an organization's AGM can be a useful indication of when to send out update forms as this is when officers or policies are likely to change.

— Other officers: names, addresses, telephone numbers of treasurer, president, chairman, publicity officer, membership secretary, *etc.*

— History of the organization including special events or persons associated with it.

— Publications: newsletter, diary, commemorative brochures, annual report, *etc.*

— Date the information was obtained and/or last updated or checked.

Agencies and Organizations Providing a Service

— *Name:* popular name and full name, *e.g.*, CHAT (Come Here and Talk), SHAC (Shelter Housing Action Centre), plus relationship to parent body where necessary.

— *Address:* street address and postal address, Post Office, box number if used.

— Telephone and fax number, including-service number, administration number, after-hours or hotline numbers.

— *Contact person:* personal name, title and address, if different from above.

— *Hours of opening:* days and times; note if seasonal, *e.g.*, holiday period, term time.

— *Services provided:* advice, counselling, practical help, *etc.*; types of enquiry the service would like to have referred to them.

— *Eligibility:* age, income, sex, residency, status.

— *Application procedures:* walk-in, appointment or waiting list; what papers or documents need to be brought.

— *Cost:* free, fees, means-tested, donations; any facilities for payments to be spread over a period; help available from local or central government, charities, *etc.*

— *Geographical area served:* neighbourhood, city, county, region, adhoc area; no geographical restrictions.

— *Branch offices:* extension bureaux, mobiles, surgeries, *etc.*; include hours of opening, routes and times of stops.

— Director, administrator or executive director of the service; name and telephone or extension.

— *Volunteers:* does the service use volunteers and for what purposes? Method of recruitment.

— *Publications:* directories, handbooks, leaflets, annual report, *etc.*

— *Funding:* local government grant-aided, donations, *etc.*

— *Special facilities:* foreign languages spoken, access and facilities for people with disabilities, photocopying or fax service, escort, advocacy, *etc.*

— *Transportation:* how to get to the service, *e.g.*, bus route numbers, underground line and station, *etc.*

— Date information was obtained and/or last updated or checked.

These two lists represent most of the facts that a community information service might need to know about an organization or service. In practice, depending on the scope of your service, it may not be necessary to include all the fields in the questionnaire. It is a good idea to get the person filling in the questionnaire-to-sign it, thereby giving the information service some-protection against claims of giving out wrong information—a not uncommon occurrence, especially when the information is committed to print.

There are other types of 'soft' information that your service might need to collect, for example halls for hire, places of local worship, accommodation, 'What's-on' events, local industry. The procedure is much the same:

— Decide what information you need to know.

— Devise a standard format or list of questions to ask.

— Identify possible sources of information.

— Decide which method or methods to use to collect the information: telephone, personal visit, postal questionnaire, press advert, *etc.*

How to Store 'Soft' Information?: There are various methods that can be used for storing information: A list is easy and quick to consult, can be photocopied for clients to take away and can be faxed, but has very little flexibility for inserting and updating information.

A strip index provides flexibility, can easily be photocopied, and is available in various forms: large binders holding many pages, smaller folders containing just a few pages, address books, wall-hung panels, rotunda units, *etc.* Strips are available in several colours, which allows for a simple categorization, and various widths. Even so, they are limited as to the amount of information that can be contained on them. It is also not quite as easy to insert new strips as the manufacturers claim.

Cards are still one of the most commonly used systems for storing information. They are infinitely flexible, easy to insert and update, can be sorted into any preferred order, and require very little technical know-how. The most popular sizes are 5in x 3in, used mainly for indexes or temporary information waiting to be checked, 6in x 4in, and 8in x 5in, which are large enough to contain sufficient information about most organizations or services and to be pre-printed with a standard grid for recording the information.

There are more superior types of card files available, such as the Rondofile or Rotadex systems which use custom- designed stationery and therefore tend to be more expensive. With systems that use standard cards, you can-if you are hard-up or want to save trees-use the reverse of scrap cards. Card files have a number of disadvantages: they are not so easy to carry around, they arc not easily reproducible if someone wants a copy of a particular section, multiple access (*e.g.*, by name, subject, area, service, *etc.*) requires duplication of cards, and the updating process can be tedious.

You may still find around edge-punched cards which offer a primitive type of multi-access to the information on the card-through the strategic use of knitting needles or rods. Personally speaking, I would save up and get a computer.

Loose-leaf binders with one sheet for each entry, sometimes called 'sheaf' binders, have a similar flexibility to cards. They are slightly slower to add to and amend but are more portable. They are suited best for alphabetical arrangements as they are difficult to guide.

Microcomputers now have the capacity to handle immense information files that can be accessed by any number of predetermined variables or even free-text searching by keyword, depending on the software used. With a printer attached, all the information or sections of it can be printed out on demand or to provide a master for printing by conventional methods. Personal computers now begin at prices that bring them within the range of most community information services. However, there is danger in automatically seeing the computer as the best solution for maintaining your information file, as this may not always be the case. A lot depends on the size of your file and its complexity, what you want to do with the file, and to what other uses you want to put the computer. The importance of the computer in, information and advice work is such that it merits a chapter to itself.

Filing System: Whichever method is chosen for physically storing the information, except for that by microcomputer, it will be necessary to decide the best way of arranging entries so that the information can be retrieved swiftly and accurately. In order to ensure that the total resources of your information service are used to the full when answering an enquiry, it will help if the system used to organize the information file can be integrated with that for the vertical file and the book-stock. This point needs to be borne in mind now, although it will be dealt with more fully later.

The bulk of the information file is likely to comprise 'soft' information -which you have collected about organizations, clubs, societies and services and, to a lesser extent, about individuals who are a valuable source of information or advice, plus items of 'hard' information not available or conveniently available in printed form. There are several ways in which you might want to retrieve this information-by the names of the organizations, by subject (*i.e.*, area of interest), by clientele, by the type of service (*e.g.*, counselling), or by place. In practice, the 'clientele' and 'type of service' approaches are usually catered for through the subject file rather than, as separate files.

Organizations File: The master file for the whole system will usually be an alphabetically arranged sequence of entries by name of organization. It is generally recommended that the full, official name of the organization is used, with references from other forms of the name. In practice, it does not really matter as long as there are references from the alternatives to whichever form is chosen. There is a good argument for putting the main entry for an organization under the form of name by which most people refer to it, even if this differs from or is an abbreviation of the official name. This is especially so where the file is to be used by the public.

In addition to details of organizations, each entry in the master file may also contain certain 'housekeeping' details relating to the maintenance of all the information files, plus references to appropriate headings in the subject file, where relevant information can be found, and a classification number.

Subject File: Most enquiries received by a community information service are likely to be about the need for a service or activity or for help with solving a problem, rather than for information on a specific named organization. So the information you have collected about organizations, services and individuals will also need to be accessible by the services and activities they provide. There are two ways this can be achieved, by either using a subject index or having a subject file of organizations. In both, you will need to choose or adopt a set of subject headings which will adequately describe the information on file or elsewhere in the system and the interests of your clientele.

If you are compiling your own list of subject headings, choose terms that are in common use by your clientele, especially if they will be consulting the files directly, rather than the 'official' term, *e.g.*, OUT OF WORK or ON THE DOLE instead of REDUNDANT. Always refer from alternatives not used to the chosen term.

A subject index is rather like the index to a book. It is an alphabetically arranged file of subject heading cards on which reference is made to where information on that subject can be found. The cards themselves do not contain information. The subject index can contain references to organizations, individuals and other supplementary material such as pamphlets, periodical articles, audiovisual material, books, *etc.*

A subject file consists of an alphabetically arranged set of subject heading cards behind which are filed copies of the organization cards appropriate to that heading and possibly cards containing items of 'hard' information. An alternative is to arrange cards according to a classification system either devised by yourself or adopted from another service, such as the NACAB classification scheme.

Place File: Where an information service covers a number of distinct towns, villages or neighbourhoods, each having a similar range of organizations, it may be helpful to have a file that can be accessed by place. There are a number of options. You could sort the organizations file initially by place, especially if it contains few organizations whose responsibilities extend to the whole area covered by your service. Alternatively, if most of the organizations are prefixed with the name of the place (*e.g.*, EXVILLE ATHLETIC CLUB, EXVILLE COUNCIL FOR THE DISABLED), the alphabetically arranged master file will automatically bring together those from the same location, provided inverted headings have not been used.

Some arrangements will have to be made for organizations which do not conform to the pattern, either because the place-name does not feature at the beginning or because it is not contained in the name of the organization at all. In these cases, an additional card can be made out with either an inverted heading (*e.g.*, EXVILLE, ROTARY CLUB OF) or the place-name can be prefixed.

The third option is to have a separate file of organization cards arranged initially by place and then alphabetically by name of organization or by subject.

Filing Alphabetically: Filing alphabetically is not quite as simple as ABC, as any librarian will tell you. There are two recognized methods, known as 'word by word' and 'letter by letter'. In word by word, entries are initially sorted by the first word in the heading, then headings that begin with the same word are sorted by the second word, and so on. Another way to describe this method, which may be more helpful, is to treat spaces between words in a heading as an imaginary letter coming before 'a' in the alphabet, then sort letter by letter.

In letter by letter, you simply ignore any spaces in the heading and sort one letter at a time from left to right. Here is how a small group of headings would look sorted by the two methods:

Word by word	*Letter by letter*
DO SOMETHING! BOYS CLUB	'DOG AND BONE' DARTS TEAM
'DOG AND BONE' DARTS TEAM	DOGSBURY RESIDENTS'
DOG SNIFFERS ANONYMOUS	ASSOCIATION
DOGS HOMES	DOGS HOMES
DOGSBURY RESIDENTS'	DOG SNIFFERS ANONYMOUS
ASSOCIATION	DO SOMETHING! BOYS CLUB

There are several other niceties to do with filing alphabetically but we only need mention here that hyphens are treated as spaces in word-by-word sorting and numerals are spelt out as said, so that 1900 (the year) is considered for sorting as Nineteen Hundred and 1,900 as One Thousand, Nine Hundred.

File 'Housekeeping': In addition to the information needed to answer enquiries, there are other items of information to do with the maintenance of the file which may usefully be included on each entry, such as the following:

— the date the information was obtained, last updated or last checked, which will indicate not only the degree of reliability to be placed on it but also when to update;

— the date a questionnaire or update letter was sent — a check for chasing up non-returned forms;

— additional contacts or other information which does not fit into the main part of the entry;

— subject headings used in the subject file;

— place headings used in the place file, if not obvious from the name of the organization;

— feedback-comments from users of the service;

— 'tracings', *i.e.*, a list of other cards that refer to they can be traced and amended or withdrawn when necessary.

The amount of this 'housekeeping' information you need to include will depend on how sophisticated you need your file to be. The bigger the information service and the larger the file, then the more likely it is that you will need to introduce a systematic procedure for processing information. In such a case it is usual for the housekeeping information to go on the reverse of each entry, using a standard grid like the one illustrated here. It is arguable that if your information file requires this degree of sophistication, you ought seriously to consider keeping it on computer.

For a small community information service, it may be quite adequate just to note the date when the information was collected, checked or updated.

Updating: By their very nature the entries in your database will be changing continually. Hours of opening, meeting places, officers, membership fees, subscriptions, charges, *etc.*, are all susceptible to frequent change. Therefore, it is important to have a system for regularly updating each entry in the database. There are several ways of achieving this but, whichever method you adopt, it must be regular and ongoing.

Interim Updating: All the entries for organizations should be checked at least once a year to ensure they are still correct. In between times, however, new information will be brought to your attention by various means, including word of mouth, newspaper reports, organization newsletters, company reports and by direct contact with the organization itself. If the source of information is reliable, it can be substituted immediately for the out-of-date information in the database but, if not, it should be noted and a further check made to verify its accuracy. A simple way to add new information to an entry in a card file is to write it onto self-adhesive labels or slips (obtainable at most stationers) which can be stuck over the original information. Even though an entry has been updated in between times, it should still be checked formally once a year.

Annual Updating: Once a year is probably a reasonable time-span for updating your database but a shorter interval may be necessary if the information content changes frequently. You will need to decide how the annual update is to be carried out. Some organizations like to update the entire file at the same time, so that they can say the file was as accurate as possible at a particular date. However, this does create a tremendous amount of extra work for a period of several weeks and you may feel it necessary to get extra paid, work-experience or volunteer help. Another disadvantage of this method is that it captures a picture of organizations at a set moment in time. Some organizations may be just about to make changes, and your records for these would be out of date for almost a year.

Alternatively, you may decide to update on a continuous basis by using either the date the record was added to the database or the organization's annual general meeting (AGM) date as a trigger. This is easy to do if your database is kept on a computer, since a date field can be used to produce a subset of records and, by linking the database with a word processing package, a personalized standard letter can be printed out for sending to each organization. It is less easy with a card or any other kind of file. You can get little coloured metal tags to clip on to the top edge of cards, a different colour for each month, or you could use an appropriate symbol on the heading of the card, such as a coloured dot, a letter or number. Tags can make the file look a bit messy and are easily knocked off if it is in constant use; symbols require the whole file to be checked through.

It is better to update records shortly before or after each organization's AGM, since most organizations, if they are going to change their officers or constitution, generally do it once a year at the AGM. This is so because many organizations do not know their AGM date a year in advance, it is enough just to indicate the month in which the AGM is usually held. For those organizations and services which do not have an AGM, you will need to use the date the information was added as the trigger.

When writing to organizations to update their information, it is better to send them a copy of their record with a request to indicate any changes, rather than sending a blank questionnaire. This saves the secretary unnecessary time in repeating information that has not altered.

An alternative to writing to each organization is to phone, but this is time-consuming and costly, and only feasible if the number of organizations to contact is fairly small. Avoid sending out updates at times when people are likely to be away, such as during the summer. There are certain times when it is more appropriate to update other kinds of information, *e.g.*, social security benefit rates (which usually change in April and September), adult education classes or school, college and university courses which change term-wise.

When an updated entry is returned it should be checked to see if the subject or place headings still apply. Minor changes can be recorded on the master card and any other cards in subject or place files; major changes may require a new set of cards to be produced. For those with computerized databases the process is simpler as just the one record has to be updated. Don't forget to change the date the information was last checked or updated.

Importance of "Hard" Information

As well as the database, most information services will need a certain amount of 'hard' information to answer enquiries. Short, unrecorded or inadequately recorded items of 'hard' information, as we have already seen, can be incorporated into the database. However, the majority of hard information will usually be found in one or more of the vast range of print forms, starting from locally produced free broadsheets to extremely expensive multi-volumed loose-leaf reference works.

Within the scope of this book, it is not possible to go into detail about the sources of this material, since they will vary considerably according to the type and subject range of the service you are operating. Community information changes rapidly so any printed sources should be treated with care. However, you might find useful a chapter on the sources of community information that I contributed to Printed reference material and related sources of information, 3rd edn., Library Association Publishing, 1990, which concentrated on organizations and on those items which are regularly updated.

The following framework for collecting community information was originally devised by Grainne Morby, who at the time was working for the Community information Project, with some additions of my own:

Distinguish between Subject Areas: In other words, identify the main topics into which the subject scope of your information service naturally divides. Throughout this framework I will take as an example an information service for which the main area of interest is housing. It is sometimes useful to apply a consistent criterion when dividing up a subject, although obviously this would not be so easy for an information service whose scope covers as broad an area as, say, 'community information'.

However, taking housing as our example, you could decide to divide it by type of accommodation, *e.g.*, owner-occupied, private-rented, council-rented, New Town, Housing Association, institutional accommodation, tied accommodation, mobile homes, *etc.* In most cases you would probably need an additional category for subjects that cut across all or more than one category. In the case of housing, this might be 'squatting' or 'homelessness'.

Since the material you want to collect will usually be written to meet a particular need or needs, it is advisable to collect information around those needs rather than for fitting into a theoretical framework.

The kind of broad client groups that might be identified in the field of housing are landlords, tenants, owner-occupiers, squatters, transients, elderly people, people with disabilities, the homeless. Then, one could identify client groups who share a common problem or need associated with their particular type of accommodation, such as eviction, dampness, house repair, renovation, harassment, mortgages, rent arrears, planning permission, redevelopment, *etc.* Once these subject areas and client needs have been identified, it should give you clearer guidelines when you come to select material from the various sources identified in the next part of the framework.

Print Sources

Official

(a) National

- *Pre-legislation:* political party manifestos and policy documents, speeches by Ministers, the Queen's Speech at opening of Parliament, White Papers, Green Papers.
- *Legislation:* Bills, Hansard reports of debates in both Houses of Parliament, Acts, Statutory Instruments.
- *Guidance to Local Satutory Bodies:* circulars and letters from Ministers, guidance notes, reports of Inspectors.
- *Reference Books:* collections of legislation, *e.g.*, Statutes in force, Legal aid handbook.

— *Periodicals:* from government departments and agencies, *e.g.*, Department for Employment gazette.

— *Guidance to the Public:* leaflets, posters, videos and other A/V material.

— *Ombudsmen.*

(b) Regional

— Health authorities, British Rail.

— Privatized utilities: gas, water, electricity, telephone.

— Consumer watchdog bodies: Post Office Users' National Council (POUNC), Office of Gas Supply (OFGAS), Office of Electricity Regulation (OFFER), Office of Telecommunications (OfTel), Office Of Water Services (OFWAT).

(c) Local

— *Local Authority Policies and Decisions:* agendas, minutes, standing orders, bylaws and regulations.

— *Guidance to the Public:* handbooks, leaflets, *etc.* from local authorities, area and district health authorities, Community Health Councils, local offices of government agencies.

Non-official

(a) National

— Reference books and handbooks from commercial publishers.

— Voluntary organizations.

— Pressure groups, self-help groups.

— 'Umbrella' organizations, *e.g.*, National Council for Voluntary Organizations, Federation of Independent Advice Centres.

— Professional bodies, *e.g.*, The Law Society.

— Trade associations, *e.g.*, National Association of Retail Furnishers.

— Educational bodies, *e.g.*, National Extension College, Open University.

— Trade unions.

— The media.

— Practitioners in the field who publish.

(b) Local

— Claimants' unions.

— Law centres, other specialist advice centres.

— Neighbourhood and generalist advice centres, resource centres.

— Local 'umbrella' groups.

— Pressure groups, campaigning groups, action centres, Voluntary organizations.

— Local media: newspapers, community newspapers, radio, cable television.

(c) New technology

— Video-text systems: Prestel, teletext, private viewdata.

— Online computer databases, *e.g.*, Volnet

— CD-ROM, videodiscs, *etc.*

This list is not intended to be exhaustive but will give you some idea of the range of sources in which to look for information. Howard Matthews, in his book *Community Information* has drawn up a set of criteria for selecting community information materials. It is aimed particularly at library community information services but may be generally helpful for anyone setting up such a service.

1. Selection should be done by those who know the community best.
2. Material should relate to specific, local, identified needs.
3. All material should be current.
4. All material should be written at a level appropriate to its intended use.
5. Material should be concerned with the practicalities of problem solving.
6. All new material should either fill a gap, offer a new viewpoint, or promise something better.
7. Material should take into account the level of use the client will make of it.

Obtaining Publications and Other Material

Once the material you require has been identified, you will need to obtain it. Some material will be priced, in which case- you will need to order it, and some will be free.

Priced Publications

The procedure for obtaining priced publications, particularly in the field of community information, is not as simple as you might suppose. Since a lot of the material may well be fairly cheap and the small or community publishers, booksellers and library suppliers are reluctant to obtain it as the profit margin is likely to be too low for the amount of effort involved. The alternative is to order direct but this also has its difficulties as many small publishers now insist on money with order, and this can cause problems for some organizations through the need to raise a cheque or limitations on the amount that can be spent from petty cash.

A few national voluntary organizations, *e.g.*, Age Concern England, operate a standing-order system for their publications-this is one solution, although you may need to accept blanket coverage of the organization's output. In the larger cities, you may find a community or alternative bookseller who is prepared to supply this material to order.

A directory of alternative bookshops is published annually by the magazine *Radical Bookseller*. Whatever the method of ordering, it is important to keep a record of what has been placed on order and from whom to avoid duplication of titles and as a check on supply. For each title make out an order or slip. This can be either a scrap 5in x 3in card or one specially printed with a standard grid. The following indicates the kinds of information that need to be recorded:

- — title of publication-more important than author for this kind of material, since authorship is often unclear;
- — author-might be an individual or an organization;
- — price, date of publication and, if a periodical, frequency;
- — supplier-this may be a bookseller, library supplier or the address of the producing body;
- — date of order;

— source of information about the item;

— location, if you are ordering for more than one centre.

You may find it helpful to use the order cards later as the basis of a catalogue of the publications in your information centre; in which case., leave space at the top of the card for either a subject heading, classification number or filing code. After an order has been placed with a supplier, the order cards are filed, preferably by title, in one alphabetical sequence known as the 'order file'. When a publication is received, the order card is removed from the file and accompanies the publication to the next stage if it is to be used later as the catalogue card. If not, you may find it useful to file the order card in a 'publications received' file, until such time as the publication is permanently recorded in the system, after which the card can be thrown away.

Periodicals: Publications which arrive at regular intervals—weekly, fortnightly, monthly, quarterly—will not require an order card but you will need to keep a record of each title taken. This should include the following items of information:

— title of periodical;

— frequency;

— supplier;

— when subscription is due for payment;

— a grid to record receipt of each issue;

— instructions for disposal of back copies.

For those using computers, there are software packages available for periodicals management but these are aimed at organizations who take an extensive range of periodicals and such packages are not cheap. If you have a database management package on your computer, you would do better to devise your own periodicals database.

Free Material: This can vary from quite substantial loose-leaf binders plus updates to leaflets, posters and bookmarks and may be required in bulk for clients to take. Since this is likely to be a frequent type of request, it may be quicker and simpler to duplicate a standard letter to send to organizations, with space left to fill in address, title of item, format (leaflet, booklet, poster, *etc.*) and number of copies required. It is advisable to keep a record of what is requested, if only as a check on whether it has been received or not.

A lot of the sweat has been taken out of identifying free materials, and addresses from where they can be obtained, through the setting-up of Camden Libraries' Free Leaflets Information Service (FLIS). For an annual fee (E120 in 1992) subscribers receive a copy of the Frills directory of free information leaflet suppliers, a monthly batch of new and updated leaflets and amendments to the Frills directory, and details of a model system for organizing leaflets. Details of this service can be obtained from Community Information Services, Camden Leisure Services Department, St. Pancras Library, 100 Euston Road, London NWI 2AJ. A similar, but not so extensive, service is provided by the NACAB as part of its Information Service, but you would need to have approval to subscribe to the full pack.

Organizing Hard Information: We have already seen that hard information can be found in a great variety of forms. For the benefit of organizing it, the following categories can be identified:

Material in book form for use by clientele or information workers will be best arranged on shelves in the broad subject groups identified when acquiring the material or according to a classification scheme.

Ephemeral material for use by information workers is best stored in a vertical file using the same subject headings as those in the subject sequence of the information file, though probably with more subdivisions, or using a classification scheme. Alternatively, the material could be kept in file boxes on shelves using broad subject headings.

Ephemeral material for use by clientele on the premises is best kept in file boxes, ring or display binders and ideally interfiled with books. Vertical files are not recommended for public use as they tend to be a deterrent. Another way of dealing with ephemeral materials is to gather them together in packs on particular topics and, for maximum impact, to display them face outwards on sloping shelves.

Ephemeral material for the public to take away can be displayed in special leaflet dispensers, on a sloping surface or on tables. Ideally, they should be displayed in broad subject groups or by originating organization, *e.g.*, Benefits Agency, Department for Employment, but this is rarely possible because even within organizations leaflets are produced in varying sizes.

Additional Information

Many organizations produce printed material in the course of their work or activities. The most common types of material are directories or address lists of members (individuals or constituent groups), newsletters, events sheets or cards, annual reports, constitutions, posters, leaflets, funding appeals, campaigning literature, advertising brochures, balance sheets and commemorative booklets (anniversaries, *etc.*).

All this material is potentially useful as backup information to the database but there is no sure way of obtaining it. When sending out the original questionnaire to organizations, you can ask to be put on their mailing list for this type of material, if one is maintained.

Some organizations that operate on a shoe-string budget may make a small charge for their mailing list facility. Even when you are put on a mailing list, regular receipt of material may depend very much on the enthusiasm or efficiency of the secretary of an organization. Often the best solution to obtaining material is to maintain regular contact with organizations and to pester them continually. When material arrives, it should be dated, so that you know roughly how old it is, and carefully filed for future use.

The simplest method is to have an envelope or folder for each organization. Write on the outside of the envelope the name of the organization as used in your database and arrange envelopes alphabetically in boxes or a vertical file. You should also add to this file any cuttings from newspapers or magazines concerning those organizations. The file should be checked regularly, at least once a year, to remove out-of-date material. When writing to organizations for updated information for your database, ask them for any new literature.

Before throwing away the withdrawn material, check to see if it would be of use to some other organization or section of your service which collects material on local history.

The Stratification

It is not my intention to go into any great detail about the classification of community information materials. For most small information services, it will be quite adequate to arrange material by broad subject groups subdivided according to the form in which the clientelė's needs and problems are presented. For ease of labelling vertical files, book spines and boxes, a simple notation can be used based on the initial letter or letters of subjects. For example, Cambridgeshire Libraries' community information service covers 21 categories:

Ben	Benefits	Gov	Government
Bus	Business	Hea	Health
Car	Careers	Hou	Housing
Corn	Community	Law	Legal rights
Con	Consumer	Mon	Money
CR	Community relations	SC	Senior citizens
Dis	Disability	Tra	Transport
Edu	Education	Ump	Unemployment
Emp	Employment	Worn	Women
Env	Environment	You	Youth
Fam	Family		

Some libraries also include as separate categories: Death, Equal opportunities, Fuel (or Energy), Gay rights, Leisure, and Trade union rights. These categories can then be broken down into smaller subcategories as necessary. Taking the earlier example of Housing, a possible range of subcategories with a mnemonic notation might be:

A	Agricultural accommodation	Re	Rented accommodation
H	Homelessness	ReC	Council housing
I	Institutional accommodation	ReH	Housing Association housing
M	Mobile homes	ReP	Private rented
O	Owner-occupied	S	Squatting

Numerals could be introduced for more detailed subdivisions, *e.g.*, 01-Buying and selling a house; 02-Home insurance; 03-Planning applications; 04-Development; *etc.*

An effective method that can be used where you have only a small number of broad categories (no more than ten and preferably less) is colour coding. Allocate a colour to each subject and attach an appropriate coloured sticker or tape to the spine of each book or file box. Colour coding can be used in conjunction with notation to provide more detailed subdivision of a subject. When a community information service grows to a size where broad categories are not adequate to 'contain all the material and allow efficient retrieval, then it may be necessary to consider using a classification scheme.

Two choices are open to you, either using or adapting a ready-made scheme or constructing your own. There are a number of general classification schemes used in libraries, of which the most well known is the Dewey Decimal Classification. These schemes have been constructed to organize the whole of knowledge and subjects are usually arranged on philosophical or logical principles. Consequently, they may not be sufficiently detailed or treat subjects in the way your clientele express their information needs. Most schemes have a degree of flexibility and it might be possible to adapt them. The NACAB has its own scheme for arranging materials in the broad area of community information, but it hardly merits the description of a classification scheme. However, it is sufficiently detailed and client oriented to be worthy of consideration.

In so far as control over information forms the basis of managerial authority, the technologies which process it are inherently political. Indeed, business organizations are under pressure to change their decision-making structures as a direct consequence of information technology implementation. These 'subversive' effects on organizational structures are not immediately obvious to management. They arise as the new technology changes the established work practices and consequently creates different relationships on the organization's information network. Management is then faced with a dual threat:

1. A new pool of knowledge outside its formal control structures. Employees are able to access information without hindrance.
2. Power is shifting from line managers to those who control the information technology, such as the technicians in the DP departments or knowledgeable users.

Management's response has been weak and incremental. The control of information generated by computer technologies increasingly reflected the new power structures that these technologies established. On the one hand, functional control (allocation, budget, *etc.*) remains with general management. On the other, operational control (how and what is processed) is left to the technicians in the DP departments. The pervasive nature of the technology spreading through all levels of the organization established those who controlled it as important powerbrokers or gatekeepers. Newcastle Business School research has shown that senior managers are often unwilling to become involved. They served their apprenticeship in computerless or, at best, early mainframe-driven organizations and therefore do not always fully understand or appreciate the new technologies. Despite the obvious need for new control structures the traditional methods are adhered to.

The dichotomy between the established management structures and the changed (and changing) technological structures is therefore maintained and often worsened. Such opposing forces create a complex environment and place considerable constraints on management action. The dichotomy quickly becomes institutionalized. Lack of management action causes other user employees to perceive themselves as powerless in any dealings with the new technologies. This serves to enhance the influence of the technicians and is built into the organization's structures. Business action then tends to become increasingly inflexible; user managers are reluctant to take risks in an environment they do not understand, and many operational decisions are left to the technicians. This area of decision-taking is also built into the firm's task structure. Meanwhile the new networked systems demand increasing flexibility in order to work but are instead being faced with increasingly inflexible and decision-shy managements.

To appreciate the nature of this threat and how profound its impact can be on the organization, it is essential to grasp the characteristics of technology, in particular its integrative nature with any social entity whether it be an organization or something else. Explore this area, but it also then distinguishes between different types of technology, highlighting the perceptual characteristics of information technology. It concludes that to control information technology we must realize that we cannot treat it as just another piece of machinery.

The relationship between technology and an organization has, indeed, been thoroughly explored. It is obvious that a firm's structure is characterized by the many forces emanating from its operating environment. What is perhaps not so obvious is that technology is one of those forces. This is because it is quite often seen as a locally owned, physical resource. The machine on the shop-floor is just that and not part of a more complex social process. Within the framework of this perspective technology is a tool, totally predictable and controllable by management.

The physical technology alone (*e.g.*, machinery or computers) will not explain how organizations are affected by it. As Weizenbaum suggested, a computer or telecommunication system (or any form of technology) is merely junk without the people to relate to them. It is how those people relate to their technology, how they use it, their perceptions of it, and so on, that are important. These aspects govern not only the technological process but also the characteristics of the organization structure. In the same way, structure should not be seen merely in a physical sense. The dividing of a firm into departments and functions, although of interest, does not capture the essence of an organization, which is to be found within the minds of its employees. The relationship between technology and structure is thus not physical but perceptual because it is within people's minds that the links are made. Our understanding of such techno behavioural forces is not great. This is partly because of the hegemonic view of business organizations as hierarchies in organization theory. It would seem, therefore, that we have not as yet learnt to examine organizations in an appropriate dimension which can highlight the effects of information and its technology.

The traditionalist's view of an organization is typically presented by Schein as a rational coordination of activities. This emphasizes the hierarchical and functionally-bound aspects. Butler, on the other hand, discussed the lack of understanding of such models determining that all the interrelationships are not adequately identified. They tend to portray firms as static entities without the turbulence of opposing forces. In particular, technology is seen as a side issue, if at all, and not central to the fundamental mechanisms of an organization.

A more appropriate model in Butler's mind is Udy's, which puts technology at the centre of an interactive environment of production objectives, social setting, physical exigencies and work organization. Technology, not the structure, sets the constraints ('trade-offs' as Butler calls them) upon an organization. Thus, a certain structure can be considered suitable for a given technology. For example, a company using mass production techniques, such as Ford, will best be structured to suit the needs of mass production technology. The relationship does not end there, however, as in Pugh's model, because he suggests that it is technology which drives the organization.

Where this model fails is in not identifying with sufficient clarity the perceptual aspects of an organization's technology. It is only within this domain that useful links can be established between technology and structure. The perceptual aspects of structure must also be emphasized for this to be successful since within any physical domain technology can only be seen as a resource and structure as a functional framework. It will then demonstrate that employees' perception of their structure determines the consequent structural characteristics. This chapter also suggests, however, that the relationship is two-way, in that an established structure can itself influence and change the perception of the employees. The impact of information technology is then again introduced. Already it has been established that such technology can change the way that individuals perceive their environments. A link with the present chapter is that since an employee's perception of the environment can be changed through information technology, his/her perception of the firm's structure must also change. The post-implementation structure that emerges may be profoundly different from any previous structure.

This control-theoretic view of information technology's impact is particularly relevant to the devolution of management authority. The impact of the new technology may radically affect the power structure or the political relationships between management and staff (but not the functional relationships) thus creating an imbalance between the perceptual and physical aspects of the organization. Through the perceptual domain, information technology has also infiltrated the cultural base of organizations and has thus had the potential to alter cultural objectives, which are the consequence of existing value systems.

For example, it may be the formal objective of one firm to produce good quality products at the expense of higher profits. This objective has arisen from the organization's culture, but any form of technology, especially information technology, is not sensitive to cultural requirements.

Managers may try to fit the technology into their companies but it is generally within a framework provided by the physical domain. The objectives suitable for the technology do not always match existing cultural needs. To seek an explanation of the technological impact, therefore, we should no longer look at organizations in terms of departments and divisions. Any boundary established by such a definition may be rendered irrelevant. It is more sensible to consider an organization as a network of perceptual maps, dependent on an information network which is itself part of a wider network within the region, market, industry, and so on.

The need for such an outlook is underlined by the number of firms accessing information beyond their established functional boundaries. Internally, this involves different departments accessing a common pool from the company database; externally, employees gain entrance to public databases such as Prestel. It is increasingly difficult to understand these organizations as physically controllable entities. The problem, however, extends beyond management. If it is accepted that our models of organization are also too limiting because of their inability to highlight the dichotomy between the perceptual and the physical, then we must develop new models, which is one of the major proposals of the book. In order to do this we must first establish an appropriate methodology. The analysis which must be pursued is wide-ranging. It extends beyond the convenient boundaries that traditional theory has established. The tidy logic of an organization reaches only into the physical domain, beyond that is a vast perceptual arena which can be understood only as a network of relationships.

Dynamic Nature of Communication

Before its presentation, however, there are other factors which must be addressed. The point has already been made that an organization is a fluid and ever-changing entity. The components so far discussed form the framework of the organization but they are no more than boxes in an analyst's model. Other forces exist which breathe life into these concepts. They can be termed as the dynamics of an organization. People are the organization, and as such their perceptions, aspirations, alliances and conflicts form an important dynamic. It suggests that the outcome for any organization is not through a rational process, although that may seem to be apparent within the physical domain, but rather as a consequence of perceptual interaction between individuals or groups. These can be dysfunctional as well as functional.

For managers the issue is how such diverse forces can be controlled. They can only directly access the physical domain and therefore have to attempt to manipulate this in some way to have any impact upon the perceptual domain. One area, however, which allows a controllable linkage between the two domains is the information network. The quality, quantity and content of information, amongst other factors such as historicity, will determine an individual or group perception. On the other hand, the network itself can be physically manipulated thus affecting the quality, quantity and content of information. Its control is therefore of considerable importance to management. Its particular emphasis is on the use of computer technology to generate information. Other forms of generated information through manual methods can also be controlled as information management, but because of the computer's ability to generate so much information other important forces come into play which are not present in a manual system.

The ways in which employees regard the characteristics of their information will determine in part the nature of their information network and ultimately the organization structure itself. Management must be aware of this and develop strategies through training and computer technology implementation which are sensitive to this. Information, however, is only part of the process. On its own it is not particularly a dynamic of the organization. Individuals may receive information and do nothing with it. A firm, therefore, does not change because of information. Information must be used. Action or behaviour is the consequence of information being used in the physical domain. In the perceptual domain, information being used is manifest through the decision process. There are no actions without decisions, no matter how minor or simple they may be, so that an organization, which can be analyzed in terms of its actions, is driven by its decision processes.

In any context there is no such thing as an individual decision, even if the person involved is on a deserted island. Especially within a complex social process such as an organization, a decision is based upon a range of perceptual and physical influences. A person's experience coming to bear in a decision, for example, is the consequence of previous interactions with other people and institutions. Therefore one cannot fully control the decision process by controlling those individuals who are seen to own it. Their actions might be constrained but the source of their decisions is to be found in the perceptual domain. This is particularly important when such processes are affected by computer technology. Managers tend to develop mechanisms around the physical aspects of the technology and ignore the wider perceptual impact. It then places these ideas in the context of the other major components of a firm, in particular the information network, by highlighting the close relationship between a decision and information.

We have so far developed the framework of an organizational structure dominated by two domains: the physical and the perceptual. In a similar fashion to matter and anti-matter, their forces to Conflict and contradict one another. Management must maintain the narrow ground between the two, because it is only here that long-term stability is established. At this point the forces of both domains are in equilibrium, which for the purposes of the model has been termed balance. The achievement of such a state is one of the key outcomes to be identified.

The domination of one domain over another could lead to long-term instability, the degree of which depends upon a range of factors unique to specific organizations. For example, if there is no balance and an organization is dominated by its physical domain, then its short-term survival, at least, will depend upon how effective that domain is in dealing with the firm's particular environment. In the longer term, however, all firms must achieve an equilibrium whether the dominant domain is in tune with the environment or not. This is because the cost of not achieving equilibrium will be manifest in the dysfunctional relationship between the two domains. Forces will come into play that will undermine stability. It shows a complex social entity driven by an abstract relationship between two sets of forces. Much of this arena exists within the minds of the employees, of course, but like other societal phenomena its logic extends beyond them.

The forces can exist independently of particular individuals in that they are perpetuated by culture, technology and training. Individuals may come and go, some people may contribute more than others, but the logic of the organization as a separate entity remains consistent. It is proposed that an, organization can pass a significant point, the virtuality point, whereby its survival does not depend upon a particular set of individuals. This is a different world from the one portrayed by established theory. It comprises many of the same elements, such as structure and technology, recognized by conventional wisdom, but it is the perspective, the way in which those elements are arranged, that is different.

The emphasis is upon a force field relationship rather than the aspects of one domain or another. By having this different framework, we can also analyze the effects of change in other ways. In particular, the impact of information technology is shown to have a far wider effect upon organizations than the traditional models may lead us to suppose. Much of it is common sense, gleaned from the working environment and from those in it. Nevertheless, the model is placed before all management as a challenge: to use the ideas this book presents as a basis for analysing the worth of their information technology.

Significant Aspects

While mankind has developed myriad ways of applying and controlling power to dominate and shape our environment, through the use of—tools, weapons, machines, fuels, vehicles, instruments, clothing, buildings and roads, metals, plastics and drugs, agriculture, and electricity—the handling of information has lagged considerably, perhaps because the human brain is itself so remarkably powerful.

Until recently, there have been only three major developments in this area: the invention of written (or painted or carved) language, some five or six thousand years ago; that of simple arithmetic operations, using what would now be called a digital representation of numbers, about a thousand years later; and that of printing, about five hundred years ago. With written language, we get the capacity to make a permanent record of information and also to convey messages across space and time: storage, retrieval, and communication.

With digital arithmetic, we get the ability to perform accurate, repeatable manipulations of quantitative data. With printing, we can make many identical copies of the same record and so broadcast a single message to a wide and continuing audience.

Beyond these outstanding advances, until the last hundred years or so, the only progress has been in the engineering department with increasingly plentiful production of more powerful and reliable and efficient, faster and cheaper devices to implement these concepts. In the last hundred years, we have seen the rapidly accelerating advent of a technology so powerful, novel, widespread, and influential that we may indeed call it the Second Industrial Revolution.

Its basis is electromagnetic, in many interconnected forms: photography, photocopying, cinematography, and holography; telegraphy, telephony, radio communication, radar, sonar, and telemetry; sound and video recording and reproduction; vacuum tubes, transistors, printed circuits, masers, lasers, fibre optics, and (in rapid succession) integrated circuits (IC), large-scale integration (LSI), and very large-scale integration (VLSI) of circuitry on a tiny semi-conducting 'chip'; and, finally, the bewildering variety of electronic digital computers.

Role of Technology

The progress has been truly amazing. In only about 40 years, electronic communications and news media have become commonplace and indispensable. Computers have proliferated, becoming increasingly fast, powerful, small, and cheap, so that now there is scarcely a human activity in which they are not to be found, bearing an increasing share of the burden of repetitive information processing, just as the machines of the First Industrial Revolution have taken over the majority of heavy and unpleasant physical labour. Now, information can not only be stored, retrieved, communicated, and broadcast in enormous quantities and at phenomenal speeds; but it can also be rearranged, selected, marshalled, and transformed.

Until recently, these activities were the sole domain of the human brain. While creative, judicious, moral, and aesthetic choices are still best left to people, all the tedious and mechanical mental processes can now be relegated to the accurate, fast, and tireless machines. Any sequence of operations on information that can be precisely specified can be carried out without further human intervention or supervision.

At first, computers were the experimental toys of university researchers; then they became the tools of government establishments and giant corporations, huge, expensive, individually designed and manufactured, and beyond the reach of any but the wealthiest organizations.

People thought of the future in terms of machines of ever-greater speed and capacity; centralized behemoths would hold all the world's information in gigantic data banks, whence major decisions would issue to be imposed upon the populations at their mercy.

With the emergence of powerful, cheap, mass-produced computers-on-a-chip, the picture has changed radically. Now we see tiny computers everywhere: in wrist-watches, microwave ovens, electronic games, pocket calculators, cameras, typewriters, musical instruments, *etc.*

What used to be done, with few options, by intricate mechanical devices is now performed, with great flexibility and convenience and at much less expense, by the ubiquitous preprogrammed microcomputer.

The probable future has become one of millions of small yet powerful computers, controlling virtually every machine and appliance. These are distributed in every homè, on every desk, in every workshop; many of them connected in a maze of small and large networks, much like the present telephone network. This enables individual computers to communicate, sharing information in a gigantic distributed database, and gaining, through distributed processing, computational power whose extent is yet difficult to gauge; all this following the individual requirements and choices of the owner or operator of each machine.

Increasingly, we are confronted, not only with the results of the use of computers throughout industry, commerce, banking, advertising, science, the communications industry, newspapers, airlines, and hospitals; but with the realistic possibility of purchasing computer power for our own small enterprises, offices, and homes. This may be done in a variety of ways; but in all of them, the real cost of computation is constantly diminishing.

It is probably fair to say that the question of computerization is not "whether," but "when" and "how." We must choose whether to lease equipment or to buy it; whether to install terminals, connected to a computerized "service bureau," or a complete local computer system; whether to get a package of standard programmes directed towards our kind of work, to commission new programmes tailored to our special needs, or to learn programming and write our own; whether to go it alone or to share a system with a number of similar users (especially if they are in the same building); how far to take our first efforts at computerization; what to include and what to continue doing by hand.

Computer programming is likely to become the literacy medium of the third millennium AD.

Elementary schools may well be teaching it before long, and we might be well advised to gain at least a smattering of knowledge of computers and of programming competence, especially since computer languages and programming environments are becoming increasingly helpful and friendly to the uninitiated user.

A computer is a machine for the automatic processing of information. Historically, this information was numerical, and computers were machines for doing arithmetic.

Unlike the simpler calculating machines, which can perform only one elementary arithmetic operation at a time, they need to be told what to do next (usually by suitable button-pushes); computers can be given a list of operations to perform (often with branching and repetitions, depending on tests of sign or value included among the operations), which they will then execute in proper sequence without further intervention. This sequence of instructions is called a *programme*.

A digital computer stores its information in the form of words, finite ordered sets of digits, each of which can have only one of a finite set of values.

Considerations of simplicity, reliability and economy dictate that electrical engineers should design computers to consist of a great number of similar pieces of circuitry, each of which can only be in one of two states, usually denoted by 0 and 1.

Such *binary digits* (or bits) are the elements of which computer digital representation is built. A row of eight bits is called a *byte,* and the majority of computers have their storage of information organized in words of one, two, four, or eight bytes (8, 16, 32, or 64 bits).

The number of bits in a word is termed its *length;* if this is *k,* then the number of possible distinct pieces of information that can be stored in such a word is 2. In particular, four bits together can have 16 different contents, and these are standard binary representations of the numbers 0-15:

0000 = 0	0001 = 1	0010 = 2	0011 = 3
0100 = 4	0101 = 5	0110 = 6	0111 = 7
1000 = 8	1001 = 9	1010 = A	1011 = B
1100 = C	1101 = D	1110 = E	1111 = F

with A = 10, B = 11, C = 12, D = 13, E = 14, and F = 15. These may now be viewed as the 16 possible digits of a representation (the hexadecimal, or hex), which is much more compact and humanly intelligible than a long string of zeroes and ones. For example, the byte 10110010 becomes "132," and the four-byte computer word 01001100011100101101000110001110 becomes the eight-digit hex word "4C72D18E."

The following are various parts of which any computer is composed. These are, essentially,

1. a *central processing unit* (CPU), which is the controlling and computing centre of the machine;
2. a *memory,* possibly of different levels, in which both *data* and *instructions* are stored;
3. a variety of *input* and *output (I/O) devices,* through which the machine communicates with the world outside it.

The CPU consists of an operation control unit (OCU), an arithmetic/logical unit (ALU), and a relatively small, very-fast accessible local memory (LM). The OCU keeps track of the memory location of the next instruction to be executed, and analyzes the current instruction, so as to activate the proper operation of a memory transfer, non-sequential jump (by appropriately changing the address of the next instruction), input or output of information, or computation (performed by the ALU), as is indicated by the instruction code.

The ALU actually carries out the elementary arithmetic operations (addition, subtraction or negation, multiplication, division or reciprocation) and logical operations (these being bit-by-bit operations, without carry' such as not, and, xor; eg., not 1010= 0010, 1100 and 0101 =0100, 0110 xor 1010 = 1100) on the data given to it by the OCU. The LM receives the operands called for by the OCU and also the results of the ALU'S operations upon them. For example, the OCU may retrieve the factors of a multiplication from the main memory into a pair Of LM registers and instruct the ALU to multiply them and place their product in another LM register. Such registers are usually called accumulators, and they are normally double-length (since the product of two k-bit numbers is a 2k-bit number). Other LM registers are used for counting (eg. repetitions) and are called index registers; others hold intermediate values and are called buffers; and, finally, there are one-bit or two-bit registers which reflect the information on which tests are made by the OCU (for example, the occurrence of a carry, the vanishing or positivity of an answer, or the occurrence of an arithmetic overflow): these are termed flags.

Originally, the CPU was a sizable piece of electronics, hand-assembled and highly complex. With the advent of micro-miniaturisation of circuitry, printing, and photographic techniques, and the mass production of components only the largest computers (mainframes) are built in the old way. Smaller systems generally have the entire CPU on a single chip. Among these, the name microcomputer is now applied to those with less than a million words of memory and a word length of one or two bytes; the name minicomputer applies to the larger machines, with two to four-byte words and one to a hundred million words of memory. (The smallest micro is probably more powerful than the big computers used by universities and industry in the 1950s.)

The *main memory* (mm) consists of magnetic or electronic components which store the information (both data and instructions) needed by the computer. The individual words are directly addressable from the CPU by number (rather like houses in a street), and their contents are retrievable in very short times, of the order of the operation time of the CPU (ranging from fractions of a *nanosecond,* 10^{-9} or one billionth of a second, for the fastest main-frames to several *microseconds,* 10^{-6} or millionths of a second, for the slower micros). This is often referred to as high-speed storage or random-access memory (RAM). While most of the mm is erasable and may be changed at will, some memory is used to store constants and often-used utility programmes and is not erasable by the CPU: such memory is called *read-only memory* (ROM). Sometimes this is optional and can be plugged into the computer: this is called *firmware.*

Note that a computer with one-byte-long addresses can have at most 28= 256 words of mm; a two-byte address can reach 256^2 = 65536 words; a three-byte address can select any of 16777216 words of mm; and so on. In practice, most micros and minis have mm ranging in size from 2 14= 16384 to 218= 262144 words. It should be noted that 10^3= 1000 and 2^{10}= 1024. Because computers are so heavily slanted towards base-2 representation, it has become almost universal computer parlance to use the prefix *kilo* or K (which usually denotes a thousand units) to denote 1024 and the prefix *mega* or M (which usually denotes a million units) to denote 2^{20} = 1024^2 = 1048576. Thus, we write 16K for 2U and V4M for 2^{18}. Perhaps the commonest size of MM consists of 64K = 65536 words.

Almost all computer instructions comprise an operation code (usually one byte long, allowing 256 possible operations to be specified), followed by an operand reference (number, index, or address) of variable length (since some operations require more data than others; for instance, the STOP instruction needs no operand, so it is one byte long).

The *extended memory* (EM) is sometimes considered under I/O devices, both because it is often physically located outside the computer (while the CPU and the mm are usually in the same enclosure), and because its speed of access is much slower than the speed of operation of the CPU and is comparable with the range of speeds of I/O devices. Most read/write memory is magnetic (optical memory is read-only, and fast RAM is either magnetic or electronic), either in the form of tape, drum, or disk, coated with magnetic material, much like sound recording equipment; it is similarly erased, read, and recorded upon by "heads" which contain electromagnetic sensors/appraisers. The cheapest (and most limited, in both speed and capacity) device is a common portable cassette recorder, with 1/4" tape cassettes. From this, one moves to specially engineered cassettes and recorders, and thence to high-speed 1/2" or I" reel-to-reel drives carrying thousands of feet of tape at very high speeds. Access times can be quite good for sequential access, along the tape, but random access time is poor at best, running to seconds or even minutes. Economy and a virtually unlimited total storage capacity (on numerous cassettes or reels; but only as many units as one has on-line tape-drives are actually accessible without human intervention) are the only advantages.

Importance of Disk Memory

When we wish for practically useful EM, combining large capacity with relative economy and speed of random access, we must turn to drum or disk memory; and, nowadays, the former have been practically replaced by the latter. Disk memory is of two types: floppy disk and hard disk, the first being the cheaper, slower, smaller-capacity option. Floppy disks are flexible, have diameters of 51/4" or 8", generally, and are removable from the disk-drive, so allowing one to build up an unlimited library of stored data. The information is stored on concentric circular tracks (not on a single spiral track, as on a sound record), on one or both sides of the disk. The number of tracks and the number of bytes per track vary (the density increasing with precision of engineering, and so with cost of the drive), but the total capacity of a floppy disk is in the range of 50KB to 1MB. The disks rotate at, typically, 300 rpm, and access time is governed by the time required to place the movable head on the right track, a fraction of a second, plus the fifth of a second taken by the head to traverse the circumference of the track, in search of a record; thereafter, consecutive bytes are accessed at some thousands per second.

Hard disks are rigid and have larger diameters. There are drives which range from anything from one to a dozen disks, rotating at about ten times the speed of floppy-disk drives (and so diminishing the access time of records in a track) with one or several heads.

Fixed-head drives naturally must have a head for each track (which costs more), but save head-movement time in random access. Winchester disks are movable-head drives with sealed-in disks, where the heads ride very close to the disk, cushioned by the layer of air between. In floppy-disk drives, the head actually rides on the disk, eventually wearing it out.

The capacity of hard-disk drive ranges from 10 MB to 100 MB in a single drive. Some movable-head hard-disk drives have removable disks or disk-packs, allowing for greater library storage.

The I/O devices are the computer's link with the outside world. In large mainframe computers, we see *paper card* (as in Hollerith or "IBM" cards) and paper tape readers and punches: the perforations in the paper carry the information. Increasingly in large computers, and almost universally in small ones, the main input is from the keyboard of a terminal. This is much like a typewriter keyboard, and depressing any key sends an 8-bit code to the computer.

When the computer is waiting for input from this terminal, it reads the code and interprets it as a datum; when it is busy, either the message is lost or it is held in a "buffer" for subsequent input (this depends on how the connection is made and what the computer is made to do). It is quite common for the computer to be connected to several terminals, all competing for its attention. This is called time-sharing. The computer cycles around the terminals, look for their several inputs while dividing its CPU time among them. The main output of the computer is to the display devices of the terminals; these are either video displays (cathode ray tubes, CRT, just like the screens of black-and-white or color TV sets; indeed, simple micros sometimes use ordinary television sets as display devices) or printers (in so-called hard-copy terminals). Of course, the computer may be connected to additional video displays and printers, of different qualities, as well as to plotters, a kind of printer for drawing graphs and diagrams. Many types and speeds of printers exist. Other input devices, such as audio amplifiers (receiving signals from radio tuners, record players, *etc.*), video receivers and recorders, laser disk drives and a variety of scientific instruments, can all be classified as transducers (devices that transform physical quantities, such as position, conductivity, pressure, temperature, vibration frequency, or amplitude, into electromagnetic impulses) linked to digitizers (which convert such impulses into sequences of zero/one pulses). Output from the computer can similarly follow the reverse process, yielding visible or audible results, or the control of mechanical or electrical equipment. Thus, computers can draw pictures (often, moving pictures), make music and other sounds, and can control appliances, machinery, and whole manufacturing processes.

In order to connect remote terminals to a computer, use is often made of telephone lines; and, for this purpose one employs a device called a modem (for "Modulator/DeModulator"), which converts the computer's digital signals to and from telephone audio signals. This has a cradle shaped to hold the ear and mouth pieces of a telephone's handset. It is also possible to connect several computers in this way. This is called the formation of a computer network. Of course, computers may also be connected by cable, fibre-optics, or microwave link. It should be noted that terminals usually, and other I/O devices often, themselves contain computers of varying degrees of power and complexity; so that even a multiuser computer with a central CPU may still be seen as a kind of computer network in its own right. The computers in I/O devices such as terminals or printers are often referred to as peripheral processors (PP).

Many networks do not have a central computer at all; but are simply a collection of independent computers linked for the sharing of information and, sometimes, computing capabilities. Often, they permit the exchange of messages *(computer mail)* and the pooling of data *(distributed data-base)*. They may also share a common bank of memory, accessible to all.

Finally, since the invention of computer networks, designers have been investigating the possibilities of computers made up of an array of CPUs *(multicomputer, parallel processors, distributed processing)*. These new ideas are very powerful and farreaching: they will probably revolutionize our ideas of computers, and of their applications, in the next few years.

The various peripheral devices are connected to the CPU without examining how. In fact, this may be done in several ways. We can simply have a separate connection (or port) for each device, but this limits rather severely the number of devices that may be connected to the CPU. Another way is to have a single bus (or connection) to which any number of devices may be attached.

The information signal must then carry an appropriate address. The bus receives all signals, and individual devices (including the CPU) seek out and decode only those addressed to them. It is also possible to have a switching device, which receives addressed data and directs them to the appropriate recipient device, rather like a central post office.

The decision on what communication arrangements to adopt is made on the basis of considerations of cost, capacity, and speed. It is often the case that the several devices forming a computer or a computer network have their data coded in different ways.

It is then the job of the CPU(s) and PPs to share the work of interpreting signals into appropriate codes for each machine. This is broadly termed the problem of interfacing devices. Sometimes, the solution is to have a standard code or structure for the communications device (one meets the s-100 bus, the rs-232 serial port, the ASCII character-code, and so on).

Another interfacing problem arises from the difference in the rate at which different devices can send and receive information (this is measured by the baud rate, named after Baudot, the inventor of the first five-hole paper-tape code; one baud is one bit transferred per second; hence kilobaud, kb, and megabaud, Mb typical rates range from 200 baud to 200 kb). One solution is to send each piece of information (usually one character at a time, which takes 8 to 10 bits) only when the last has been acknowledged (this is referred to as a handshake); this is sure, but slow. Another way is to use a storage buffer in which a large batch of information is accumulated for fast transmission, thus not wasting the time of the faster device.

One last kind of choice must be mentioned: some channels of communication are serial (they transmit one bit at a time), while others are parallel (they can transmit a byte or a whole word at a time); the latter are obviously faster, more complex, and more expensive. When devices are connected by cable, the degree of parallel communication is exhibited in the width of a flat ribbon cable, carrying several wires, side by side, and in the number of pins in the plugs and sockets by which they are connected to the machines. Parallel transmission is a variation on multiplexing. What we have described is generally referred to as the hardware of a computer. By contrast, the programmes that make the computer work, the "soul" of the machine, as opposed to the hardware "body," are collectively called its software. Inevitably, there came to be programmes that were hard-wired (in the now outdated phrase) into the computer, in the form of ROM. These are termed firmware.

A computer without software is a helpless set-of circuits, and the expertise required to create the basic software that will bring the machine to useful life is comparable to that required to design the machine itself. Indeed, these days, computers are designed in cooperation between computer architects, who design what the computer will do, hardware engineers, who design how it will be constructed to be able to do it, and software engineers, who design and programme the operating system that will run the machine. Beyond this, the computer will also need application software of many kinds, to enable it to do a variety of jobs, such as file-handling, accounting, statistics, payrolls, inventories, complex graphic displays, games, and so on. Typically, the application software is written (*i.e.*, programmed) by the endusers (if they are sophisticated enough) or firms of consultants and programmers and system analysts, often called software houses. Sometimes, a software house will produce part or all of an alternative operating system, or an addition to an operating system, to make a computer more flexible or more efficient than the manufacturer's own software system allows it to be.

Different Languages of Computer

The CPU of any computer is designed to accept and execute a specific set of *operation codes (op-codes)*, ranging in number from a dozen or so to several hundred. Different makes and models of computers and microprocessors may have entirely dissimilar op-codes; but the operations that they represent are much more alike than different, both through functional necessity and historical development.

The interpretation of the op-codes is built into the hardware of the OCU (though sometimes the details of interpretation may be modified by the user through what is called *microcoding)*, and it is part of this interpretation that the complete instruction being decoded contains a certain amount of further information (such as parameters, indices, and one or more memory addresses).

The "length" of a computer instruction (the number of consecutive memory words, or of bytes, occupied by it) may vary, but the OCU automatically adjusts to this. An executable programme (sometimes called "object code") consists of a sequence of machine instructions, consecutively stored in the computer's memory and (with the exception of "jump" instructions) executed in the order in which they are stored.

The aggregate of possible machine instructions is called the machine language. In the computer, a programme consists of a long string of binary digits (bits), usually written as 0's and 1's; and, of course, the same programme would be interpreted quite differently (usually as nonsense) by any computer for which it is not designed. Indeed, the slightest error in a programme almost always leads to an error in its output (usually a fatal error!). This state of affairs is sometimes expressed by the computer adage, "Garbage in; garbage out," or just "GIGO." Errors in programmes are called bugs; and the tedious, odious process of finding and correcting such errors is termed debugging. It is estimated that, in the production of a working programme, the debugging time may be two to four times as long as the time it takes to plan and write the programme initially.

To give the reader a feeling for the nature of machine language, we present a simplified, fictitious, but typical, machine language specification. Our computer has two 16-bit accumulator registers (acc), X and Y, which may be coupled into a single 32-bit ace XY, with X holding the more and Y the less significant digits; these are attached to the ALU; and a programme control register (pc) Z, also of 16 bits, attached to the OCU, which contains the address of the next instruction to be executed. Instructions will contain a 2-bit acc code a, with a = 0 referring to X, a = 1 to Y, a = 2 to XY, and a = 3 to Z. The memory consists of 216 = 65536 16-bit words, directly addressable with a 16-bit address, n.

There are four addressing modes, denoted by a 2-bit mode code d, with d = 0 referring to absolute address (address n\0 refers to the actual number *"n")*, d = I to direct address (address n\1 refers to word n in the memory), d = *2* to indirect address *(n\2* refers to the memory address which is the content of the memory word with address *n),* and d = 3 to relative address (n\3 refers to the memory address that is the sum of *n* and the content of the pc register Z, with any carry to the seventeenth bit ignored). We will write {{*a*}} for the acc denoted by *a*, *(n\d)* for the memory word with address *n* in mode *d*, and C[x] for the content of word or acc *x* (so, *e.g.*, C[{n]2}]=C[C[n]], while C[{n\3}]=C(n+C[Z]). *A port code p* selects one of 8 *output ports* (*p=0*, 1,..., 7) and 8 *input ports* (*p = 8*, 9,..., 15), these ports transmitting 16 bits at once and having to be reactivated before each function (*i.e.*, each output instruction sends out one 16-bit number, and the machine must wait for the output to be acknowledged before making another output; while each input instruction reads one 16-bit number, if an input is available, and the machine must wait for a new input each time). Finally, a bit code b refers to each of the 16 bits in a word (bit 0 being the least significant-rightmost and bit 15 the most significant-leftmost). In some cases, the codes *a*, *d*, *p*, and *b* are interpreted somewhat differently, depending on the particular op-code *c*.

As an example of a very simple computer programme, we consider the solution to the following computer problem. Our computer is to be fed a sequence of one thousand 16-bit numbers at input port #9 (these may be keyed in by hand or fed in by a digitizer connected to some experiment). They are to be stored in memory words with addresses 5000, 5001, 5002,..., 5999. Their sum is to be computed and stored in address 6000 and output to a printer through output port #2. The programme is to be stored beginning at address 0 in the memory.

Again, what is important about this example is not its detailed form, but the difficulty of its interpretation, and therefore also the difficulty of verification and debugging.

The programmer must deal with a mass of details that are of a purely mechanical nature and have no relevance to the problem being solved. "Higher-level languages" are attempts at making the computer understand the programmer's way of thinking, rather than forcing the programmer to think like a machine.

Languages used in Computer

After some introductory description of an imaginary computer, and especially of its CPU, and the establishment of some essential notation.

Presented with a set of arithmetic and other transformations required by potential users of a proposed new computer, the electronic and logical design engineers seek the simplest circuitry that will execute operations sufficient to generate all the required transformations. Circuits must be simple to be fast, efficient, reliable, and cheap. However, when an instruction has 16 or more bits, most of which should for the sake of efficiency, have some significant effect; the exact and complete specification or the action induced by it may well be somewhat forbiddingly intricate! And indeed, simplicity of circuitry does not usually lead to simplicity of use.

An ample of a programme in this machine language, which takes 16 instructions and 28 16-bit words of memory storage, to read in one thousand numbers, store them in consecutive memory locations, sum them, and print out the sum. However, please note that the English description of what the programme does (though much easier to say and comprehend) itself requires 103 characters, each equivalent to 8 bits; so that it is almost twice as long as the machine-language programme. Indeed, the difficulty of the latter lies not in its length, but in its alien and opaque form of expression of what it does. The first improvement on machine language was assembly language. This is not very different from machine language, but removes the most glaring and trivial irritations. First, instructions may be labelled, so that one no longer needs to count lines to address jump instructions; one jumps to the label. This is also a great help when one wishes to insert additional instructions, either to correct or modify a programme (every jump need not be changed). Second, memory may be labelled, so that numbers may be stored in symbolic addresses and retrieved therefrom, leaving it to a computer programme to assign actual storage. Third, the entire programme becomes relocatable anywhere in the computer memory. Finally, and perhaps more trivially, the op-codes may be replaced by abbreviated mnemonics, and the instruction layout may be relaxed, using punctuation marks. Of course, a programme written in assembly language is no longer directly intelligible to the computer or executable by it. It becomes a piece of textual input, to a "translator programme" called the assembler. Naturally, the programme may be stored in the computer's memory, like any other piece of text; but, before it can be executed, it must be "assembled." The assembler finds appropriate space for the object code (*i.e.*, the machine language equivalent of the input programme) and for its needed storage space (often called variables). The assembly language instructions correspond, one-to-one, to machine-language instructions.

For example, we may denote memory locations by one, two, or three lower-case letters (both for labelling instructions and naming variables). A numerical address will still be allowed. Absolute addresses (*i.e.*, actual numbers; $d=0$) will correspond to quoted numbers (*e.g.*, "6000"); direct addresses ($d=1$) to variables named as stated; indirect addresses ($d=2$) to variable names placed in parentheses; and relative addresses will not be used.

With relocatable programmes and symbolic addresses, they are no longer useful). Accumulators (X, Y, XY, or Z) will not be referred to symbolically, but only by these names. Instruction labels will precede the instruction and will be terminated by a colon (:). Spaces will be ignored. Op-codes will be denoted by strings of capital letters and terminated by a comma (,), which will also separate multiple arguments.

A sample specification of assembly language op-codes is shown in Figure 1. In the assembly language notation, A denotes X ($a=0$), Y ($a= 1$), XY ($a=2$), or Z ($a=3$). Italic letters (*p*, *x*, or *b*) denote numbers (possibly in quotes, to denote absolutes) or variables names (possibly in parentheses, to denote indirect addressing), and are interpreted according to the op-code (*e.g.*, A cannot be XY in IN and OUT instructions; if A is XY in instructions with $c = 1, 2,$ and 4 (with $p > 5$), the operand must be 32 bits long-either a double-length number, or an address referring to the first of two consecutive 16-bit memory words. Again, the reader must remember that it is not the particular details of our specifications that are important or memorable but the kind and degree of detail occurring.

Assembly language mnemonic	Machine language instruction
IN A, (*p*-8)	$c = 0, a ** 2, p >8$
OUT A, *p*	$c = 0, a ** 2, p < 7$
TO A, *x*	$c = 1$

FROM A, x	$c = 2, d > 1$
FROMA, B	$c = 2, d = 0$ -
CL, A	$c = 3, d = 0$
NEG A	$c = 3, d = 1$
NOT A	$c = 3, d = 2$
RNDx	$c = 3, d = 3, a = 2$
ADD A, x	$c = 4, p = 0$
SUB A, x	$c = 4, p = 1$
AND A, x	$c = 4, p = 2$
OR A, x	$c = 4, p = 3$
XOR A, x	$c = 4, p = 4$
LADD, x	$c = 4, p = 5$
LSUB, x	$c = 4, p = 6$
LAND, x	$c = 4, p = 7$
LOR, x	$c = 4, p = 8$
LXOR, x	$c = 4, p = 9$
CSHL A, P	$c = 5, d = 0$
CSHR A, P	$c = 5, d = 1$
ASHL A, P	$c = 5, d = 2$
ASHR A, P	$c = 5, d = 3$
JMP, x	$c = 6, p = 0$
JPCZ A, x	$c = 6, p = 1$
JPCNN A, x	$c = 6, p = 2$
JPCNZ A, x	$c = 6, p = 3$
JPCP A, x	$c = 6, p = 4$
JPCB A, 0, b, x	$c = 6, p = 5$
BIT A, 0, b	$c = 6, p = 6$

BIT A, 1, b	$c = 7,\ d = 1$
BITNOT A, b	$c = 7,\ d = 2$
STOP	$c = 8$
INTER	$c = 9$

***Fig.3.1** : Typical assembly language mnemonics*

Consider Figure 3.2. Here, numbers are given in decimal notation. Note the tremendous improvement in direct legibility of the programme in comparison with the binary strings. The assembler is left to figure out where to store the sum (the variable "sum" which the machine language programme put into address 6000) and the current address of the listed numbers (the variable "w," which the machine language programme puts at location 28 if the programme begins at location 0, necessitating counting the words occupied by the programme; if the programme were to be modified or relocated, this would have to be changed, and would be an unnecessary source of programming errors); and similarly, there is no need to count words to determine the addresses of "ret" and "out" (placed by the machine language programme at 5 and 24 ' as it happens). As an exercise, the readers may wish to try to modify the programme so as to allow for the possibility that the sum of the 1000 numbers might occupy more than 16 bits. In machine language this requires no less than 12 changes and seven shifts of memory location.

```
      CL X; FROM X, SUM; TOX, '5000'
ret:  FROM X, W; SUB X '6000'; JPCZX, out
      INX, 1; FROM X, (W); ADD X, sum
      FROM X, sum
      TO X, W; ADD X, T; JP, ret
out:  TO X, sum; OUT X, 2; STOP
```

***Fig.3.2** : The summation programme in assembly language*

In assembly language only the substantive changes (from the programmer's point of view) need be made. The amended programme is given in Figure 3.3. Here we note that only seven changes are needed; but observe that "sum" becomes a double-length address-just another piece of bookkeeping handled automatically by the assembler.

CL XY; FROM XY, SUM; TOX, '5000'

ret: FROM X, W; SUB X '6000'; JPCZX, out IN X, 1; FROM X, (W);

TO XY. sum ADD XY. (w);

FROM XY, sum TO X,W; ADD X, T; JMP, ret

out: TO XY. sum; OUT X, 2; OUT Y. 2: STOP

(Underlined material represents changes from previous programme).

Alternative version would have third line replaced by:

CLX: IN Y 1, FROM Y (W); LADD, sum FROM XY sum

***Fig.3.3** : The summation programme modified for long sums*

Of course, a programme in assembly language still represents, instruction-by-instruction, a programme in machine language, and therefore still has the defects of "walking the road in extremely short steps," as we put it earlier.

All that shuffling of numbers in and out of the accumulators, counting and restricted testing, is not natural to us, and its mechanical nature suggests that an improvement is still possible. Thus it was not long before programmers devised much more humanly natural languages, which are generically called *higher-level* (or *algebraic) languages.* Some of the more common ones are ADA, ALGOL, APL, BASIC, C, COBOL, FORTRAN, LISP, PASCAL, PL/I, RPG-II, SNOBOL, SPITBOL, WATFOR, and WATFIV. There are more; and all come with a variety of versions and dialects.

The characteristic that puts a language into this class is that one higher-level instruction translates into several assembly or machine language instructions. Beyond this, the languages differ according to the kind of programme they are intended for.

While most languages will perform most tasks, FORTRAN was clearly intended to do scientific and engineering computations, COBOL to perform business data-processing, LISP to manipulate lists, and

SNOBOL, strings of text. As a final illustration of the power and intelligibility of such languages, we present higher-level versions of our summation programme in BASIC, WATFIV, and PASCAL. Once more, note that the details of the individual languages do not matter to us at this point (though, if the reader ever intends to programme, then some language will have to be thoroughly understood; but assume that many of my readers have no such intention); it is their general flavour and appearance that is noteworthy; and the casual readability of the samples presented is their salient characteristic.

Of all the hundreds of programming languages devised, it is safe to say that more lines of programme have been written in FORTRAN or COBOL than in all others combined, though this reflects their age rather than their desirability. The fast-gaining runner-up must surely be BASIC, which comes with every micro and most minicomputers; though some would say that BASIC, which perpetuates the style of FORTRAN, spoils any chances of producing really good programmes and should be replaced by PASCAL. Fashions come and go in programming languages, not always fully based on rational arguments, and personal preferences vigorously touted.

Given a higher-level language, it is necessary to have a translator programme to turn it into machine (or more frequently assembly) language. Such translators are of two kinds.

An interpreter operates at execution time: Beginning with the first instruction of the stored higher-level programme, it translates it into one or more machine language instructions, which it proceeds to execute; then it reads and translates the next higher-level instruction and executes that; and so on. If a jump is encountered in the source programme (as the higher-level language programme is often called), the interpreter goes to the next instruction in accordance with the jump and proceeds to read, translate, and execute it, without taking into account whether it has already encountered it. Thus, the same source instructions will have to be passed and translated a thousand times (a very time-consuming and wasteful procedure). But the (usually much longer) object programme need not be stored, and interpreters themselves tend to be shorter and simpler programmes. This is why a BASIC interpreter is a natural adjunct to a microcomputer, which tends to be a little cramped for space; especially since many microcomputer users are riot interested in extremely lengthy computations.

By contrast, a *compiler* is a programme that translates a source programme into an object programme, the former being in higher-level language and the latter in machine or assembly language.

The programme is translated as it sits in memory, not during execution; so that the problem of repeated translation is avoided. Indeed, there are many so-called "optimizing compilers" which pass through the compiled object code several times, eliminating redundancies and inefficiencies to produce faster object programmes (something that would be impossible for an interpreter to do). Speed is greatly increased, in general, at the expense of space in the computer memory. Of course, once a programme has been compiled and proven to be free of bugs (debugging is far easier in source code than in object code), the source programme may be stored for future use and only the object programme kept in main memory. Compilers are available for almost all higher-level languages on almost all machines, and are essential for extensive applications.

Significance of Software

The programmes, or software, available for computers fall into two classes, *system software* and *application software*. The system software comprises those programmes that are considered indispensable to the general operation of a given computer system, forming what is often termed the operating system of the computer, and generally are supplied by the manufacturer, though alternative operating systems are sometimes available from software houses. The applications software includes all those programmes that are needed by one computer user but not by another, even if such programmes are widely required, and while some manufacturers will sell applications software (as optional additions to their systems), more often this is obtained from software specialists.

We begin with the *operating system*. This may be divided into a kernel of absolutely indispensable programmes and a shell of almost indispensable, so-called utility programmes. The kernel (or nucleus) is also sometimes termed the monitor. To understand why this kernel programme is needed, we consider what happens when we type a character (say an "A") on our keyboard, as part of a message to the computer. Because our timescale recognizes perhaps 1/20 second, while the computer may deal in units of 1/1,000,000 second, if we want to type an "A" we may find ourselves actually transmitting some 50,000 A's to the computer, before our finger leaves the key! If the keyboard mechanism were to transmit the A for only one microsecond, on the other hand, the computer might will be busy elsewhere and miss it altogether. Therefore, it is necessary to establish a protocol (in the usual diplomatic sense, a formal structure for the orderly passage of information) or "handshaking" procedure. This is often done by sending the A on one line (or lines) and simultaneously sending a pulse along an auxiliary line, which sets a "flag" bit in the CPU. When the computer is ready to receive a new character, it checks the appropriate flag bit over and over, until it detects that it has been set. Then it reads the character, resets the flag to its "null" state, and sends a signal pulse back to the keyboard (or its controlling circuitry), telling it that the character has been received. The flag is always open, so the transmit pulse need only be sent once by the keyboard; but the character continues to be sent until it is acknowledged by the computer. Thus a keyboard character will neither be missed nor read repeatedly. In addition, either the CPU or the keyboard controller should send the "A" to whatever device displays the typed characters.

If the particular keyboard were to be the only input device ever to be connected to the CPU, it would be best to incorporate all this in the circuitry of the CPU; but since there may be several different inputs to the CPU, and several types of terminals may be used, coming from different manufacturers, it is found preferable to have the I/O procedures made part of the operating system. In addition to the keyboard input already described, there will be short machine language programmes for input from any other devices attached to the computer and corresponding programmes for output to display screens, printers, and so on. The orderly operation of a computer is subject to interruptions (or *interrupts,* as they are known in Computer jargon), either because a programme contains an instruction that at some point, is not executable (such as a division by zero or a reference to a nonexistent memory location), or because someone needs to terminate execution for some overriding reason (such as the decision that a programme is in error and is generating garbage or is in an infinite loop). Handling errors and interrupts is another function of the kernel of the operating system, as is the start-up (or "bootstrap") procedure which initializes operation. If, as is often the case, several users are connected to the computer, then the operating system must handle the tasks of Job-scheduling (and Job accounting and billing, if this is appropriate), and the allocation of storage, in main and extended memory, and of other resources (such as printers or communication lines), and the management of *time-sharing* (as between several terminals). Another task that is handled by the operating system is the management of *user files,* and their transfer between main memory and extended memory (such as disk). Indeed, one sees frequent reference to *disk operating systems,* such is the importance of this function of the operating system.

Another function of the kernel is to provide protection and security to users and to itself, both from authorized users encroaching on forbidden territory and from unauthorized users attempting to use the computer.

Ideally, each user should, on giving the correct access code (such as a "password")—this feature did not exist on the first computers, and is still absent in personal microcomputers with only one user/owner-who has access strictly limited to the parts of the computer allocated to him or her by the operating system, but in such a way that the impression is maintained that he or she is the only user present.

Another very congenial aspect of this concept is that some operating systems conduct their memory management function in such a way that the user need make no distinction between main and extended memory; this is referred to as a virtual memory system. The kernel programme has to be able to display, move, modify, and search, at least the main memory, and to initiate execution at any given address, or activate any of the peripheral devices. Using these functions, by means of appropriate commands in the "operating system language," together with its file-handling capabilities, the system can load, compile, or execute any programme stored in its memory (main or extended) written in machine, assembly, or higher-level language, provided that a suitable compiler or other translator is available to it.

We now turn to the utilities provided by most operating systems. First, we have an assembler (and perhaps also a *disassembler,* which translates a programme written in machine language into the more intelligible assembly language; as well as a *macroassembler,* which allows the user to define his own macro-instructions in machine or assembly language), together with a selection of translation programmes, either interpreters or compilers, for the higher-level languages that the user wishes to employ.

The smaller microcomputers will provide an interpreter for some dialect of BASIC, since this a relatively simple language to learn and to interpret into machine or assembly language; beyond this, one must pay for additional languages. There may also be a variety of TRACING and DEBUGGING utilities, according to the cost, size, and sophistication of the computer and its operating system.

There will be facilities for linking or chaining programmes together. While the kernel will contain the rudiments of a file-handling system (to create, destroy, list, locate, and transfer files), there will also be utilities for further management of these. Such programmes will sort, find (among other files), search (in a given file, for information specified), transform, edit, and combine files. Indeed, a good programme editor can enormously facilitate the rapid and painless creation and modification of files, which include both text and programmes.

The line separating the kernel from the shell of utilities is purely conceptual and far from sharp. Similarly, the boundary of the entire operating system is far from definite. What is available beyond the essentials mentioned above will be called part of any decent system by some, part of a compiled language by others, and just applications software by still others, depending on their point of view. The ability to handle a variety of data structures (such as arrays, lists, strings of characters, trees, queues, and stacks) may come from the use of a suitable higher-level language or from an extended "system development" utility package.

Similarly, languages intended for scientific and engineering applications usually handle floating-point and multiple-precision arithmetic, though this may be provided even in computer hardware. The same applies to routines for computing, for example, sines, cosines, logarithms, *etc.*, and pseudorandom numbers (the last beloved of, computer games inventors!).

A *data-base management system* (DBMS) may be just a glorified file-handling utility; or may be an elaborate programme, cross-indexed and relational, with its own language of special commands, for answering any conceivable question about a large amount of intricately structured data.

As is often the case, you get what you pay for, both in money and in memory space; and vendors' claims have to be carefully scrutinized and verified. It is advisable not to overbuy one's capabilities beyond one's needs.

Similarly, a graphics package may simply allow one to produce passable graphs, histograms (*i.e.*, bar-graphs), pie-charts, and perhaps games, with a resolution of some 200 by 300 delightfully named pixels; or it may allow you to resolve perhaps 2000 by 3000 pixels, in a variety of colours, with the ability to draw complex three-dimensional shapes, properly shaded, illuminated, and textured, and move them by commands in a special language. The latter systems are a lot of fun to operate, and can be most helpful to draughtsmen, animators, film designers, and simulator-trainer designers; but they cost a bundle, and they require quite powerful computers to hold and run them.

Another offshoot of the file handier is a cluster of programmes for generating reports, journal ledgers, and accounts of all kinds, as well as forms and mailing' lists (the latter leading to a kind of DBMS in which mailing lists are matched to interests and characteristics of the individuals itemized).

Again, there are statistical packages of various degrees of sophistication, from a mean-variance-covariance calculator, to systems able to analyze very complex sets of data by elaborate techniques, using a whole statistical computer language. What has been described as "the most popular programme ever written" is usually given the generic name of an *electronic worksheet,* though the trade name of the first such programme, Visi-Calc, like Kleenex and Vaseline, has become almost generic.

A table is presented on the video screen and each entry is allocated either a numerical value or a formula relating it to other entries. When the data are sufficient, the resulting numbers are displayed. When an entry is changed, it and all entries depending on it are altered accordingly at once. This provides a representation of a given situation surpassed only by a graph in its impact, and a first-class planning aid. Finally, among the borderline system/applications software utilities, will put what is usually called a word-processing package (though would prefer either character or text processor). This is an extension of an editing utility, in which the text may by "scrolled" up and down on the screen, edited in the usual ways, but also more specifically for producing letters, articles, reports, and other written copy.

Margins may be set and the text right, left, or double justified, or set up in multi-column pages, with page numbering, indexing, and even the use of different sizes and styles of typeface, in the most sophisticated systems. Here again, cost rises steeply, and one should buy only what one will need.

Beyond these programmes, there lies an endless variety of unquestionable applications programmes. There are programmes for ballistics, boat design, analysis of molecular structure from x-ray diffraction data, tabulation of Bessel functions, simulation of naval battles and economic cycles, *etc.*

Most of these are not in the market and only work on one machine; but there are very many programmes available in the open market, produced or distributed by software houses, with various levels of efficiency, and sophistication. *Programming* is a fascinating, intricate, rewarding, but unforgiving and at times infuriating occupation. We will encourage all of you who have the time to try it. If you have a computer at your disposal, it would be a shame not to learn a simple language, such as BASIC or PASCAL and try your hand at writing a simple programme or two. The sense of achievement when you have a working programme is great; perhaps because the process is addictive and consumes much more time than you would believe possible.

The Concept

Information is one of those misunderstood concepts. Yet it is at the same time one of the most used. We go to the bus station to seek information from timetables. We obtain information from government offices such as the DSS and the DVLC. Banks, supermarkets, leisure centres, libraries and even the police ply us with more. Newspapers, television and the radio present us with their own ideas of what information should be. In many instances, however, we may not agree with their conclusions. McLeod termed its subjectivity as being one person's junk and another's treasure.

We all understand and deal with this accordingly when extracting news from the media, for example. Yet many of our organizations consider information to be something more. They see it as a vital resource, to be managed like any other valuable resource. How it is used and disseminated through the available technology can determine how efficient, and indeed effective, an organization is. Peter Drucker sees information acquired in a systematic and purposeful way as enhancing an organization's productivity.

Information is important, we cannot operate without it. But more than this, we are discovering that our ability to process it by increasingly sophisticated technological means is fundamentally changing the way that employees perceive their organizational environment. The consequences of this could be either to break down established functional controls or indeed to enhance them by becoming super-efficient. Much will depend upon the characteristics of the organization before implementation.

As firms become increasingly 'information dependent' we need to identify the perceptual relationship between ourselves and the information we use. How, for instance, can we possibly design an adequate information system if we do not understand the nature of information? The answer to this lies, in part, with the conventional wisdom of the day. Managers, like everybody else, develop their views through exposure to established ideas. Thus, the way in which they understand information will influence the way in which they treat it. Such conventional wisdom is coloured by the technicians, on the one hand, and the user's own experience on the other.

Anecdotal prescriptions abound as guidelines to the way presented for all who care to listen. These arise from a blanket of professional and academic thinking which surrounds the business environment and provides remedies for action. We could categorize all these ideas into two bodies of thought, two paradigms. The first could be termed the resource-driven paradigm. This is because its central theme in understanding information is the continuity and consistency of the information itself. It is very much in vogue at present.

The second body of thought is the perception-driven paradigm. Information is seen as an abstract concept, the product of individual perception. It is a temporary phenomenon and as such belongs only to the receiver. The difference is not merely one of academic debate. Managements adopting one or the other can affect the design of their organizations. If information is considered to be a resource then resulting systems are usually more centrally controlled, the assumption being that all information is corporate property. Whereas information considered to be personally owned is seen as being outside the formal structure.

Within the framework of this paradigm the view of information is coloured by its use as a resource. Like any other resource it can be tapped at any time with the certainty of achieving a predictable value from it. Information is regarded as unchanging, and therefore can be easily accommodated into a firm's formal procedure. There are a range of propositions available which seeks to explain information. Each proposition has a consequence for organizational design. Listed below are those major themes and what their implications might be on business.

Herbert A. Simon took the view that information, along with energy, constitute the two basic currencies of our society. We need energy to breathe, to move, to think, to live. But this alone is not enough. We need to know when to breathe, when to move and how to think. Information provides us with that knowledge. To adopt this perspective is to interpret information as an independent entity. It is all about us waiting to be picked up and used. There may be minor variations in its interpretation by different individuals, but consistency can be easily achieved through better training.

Specific sets of information, such as finance information, are given a unique value by organizations which are usually based upon departmental rather than individual needs. Particular information is thus attached to particular departments. This allows information to be generally accessible because perceptions of it are bound by the formal departmental framework. Information is then tied into that framework and cannot be used legitimately within any other context, thus guarding its consistency. What is excluded, therefore, is an important role for the individual receiver. This is not to suggest that information cannot be perceived by individuals. Psychologists such as Anderson and Bower have long been aware of the importance of information reception in determining an individual's behaviour.

However, information can be received automatically without an individual having to understand it. For example, through training an employee could process production statistics without knowing what they mean. In this case it is the functional process (through its formal procedure) that is acting as the receiver rather than the individual. An accounting process, for instance, responds to specific inputs without anyone having to understand them. A motor car responds to certain information transmitted through mechanisms such as the throttle, steering wheel or brake. Both of these examples are designed to act as receivers to specific information.

By implication, information can be transmitted to any receiver that logs into the particular transmission. In the same way as a radio picks up a station when it is switched on, the right form will pick up the right information. And like radio waves, information is omnipresent, only needing the correct tuning. Such a view allows information the continuity and consistency required to be regarded as a resource. It also allows a further proposition.

The design of any formal information system must assume that information does not change during transmission. It can, however, be transformed from data or other information sets before transmission. Within such a framework it is the transmitting functions (departments, processes, *etc.*) which are considered to be the main determinants. The information receiving functions (also departments, processes, *etc.*) are designed to link into the operational needs of the transmitter functions.

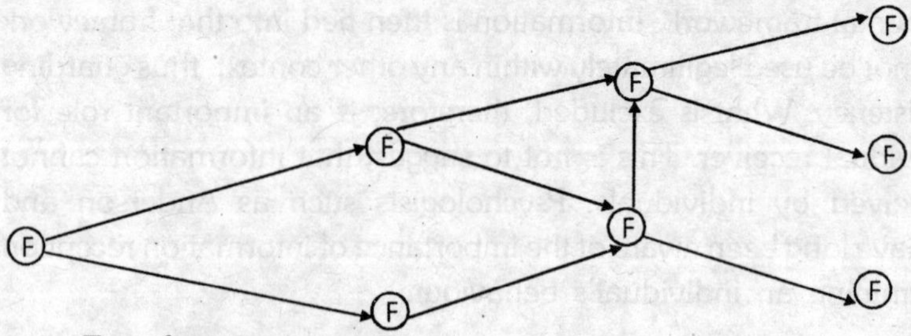

F = functions

***Fig.3.4** : An information network between functions*

Each function, however, can be both transmitter and receiver of information, and together with other functions will form part of an integrated network (Figure 1). The functions are tied by their simultaneous roles as information receivers and transmitters, thereby being prevented as much as possible from deviancy in their information usage.

Consequently, business organizations are driven by those functions which transmit information. The higher the value that a particular set of information is given, the more important will be its relating transmitter function. For example, the finance department could both transmit and receive information. Indeed, each job function within that department could also do the same. A cost clerk, for instance, receives information from the factory floor which he then processes into other information and transmits to some other source, perhaps management. The value of the information he processes will depend upon the utility his organization has placed upon it. If the utility is high, then the prestige attached to the transmitting function will also be high irrespective of job complexity. In the same way, if a department has a great number of important information transmitters within its control, then it too will be prestigious and possess influence. Why certain sets of information are prestigious and others are not will depend, amongst other things, upon the organization's political processes.

Model of Perception

The perception-driven paradigm does not consider information to be a resource. Individuals or groups are seen to own their information. What belongs to the organization are the data sets, the facts and figures specific to certain functions. For instance, data on how an organization is performing in its market quite obviously belong to that firm. How well that data is used is dependent upon the individual's own competence in interpreting it. Management can control this indirectly through training, but they cannot directly control the thought processes which turns data into information. Once again, there is a collection of propositions or ideas about information which formulate a general approach.

Adherence to this paradigm, as in the previous paradigm, may not necessarily be by conviction but rather by default because people not treating information as a resource automatically form part of this group. However, there are others who are members by conviction. Information is said to be receiver-dependent, in that, any set of information can be considered to be information only if it is recognized as having value by a particular individual. Thus, a river flooding may be information to someone who lives on its banks but of no consequence to someone living several miles away.

It would be difficult, therefore, to understand information as a group or organizations property since its determination is subjective. Management teams have implemented computer technology and in so doing also developed their information sets into a resource. They experience problems, usually some time after a successful launch, and then abandon using their technology in an integrative manner, returning to the 'islands' approach (that is, a computer system, perhaps physically integrated through networks, *etc.*, but perceptually segregated into different functional areas). Information is no longer seen as a resource; indeed its treatment as such is often seen as the culprit and not the technology itself.

Both paradigms recognize a difference between data and information. Computer personnel, for example, have long known of the distinction. Information is seen by them as processed data. The two are interrelated, data being the input to a process, and information the output. In the same way that an output of one system can form the input of another, information from one system can form the data of another. This is consistent with treating information as a resource. Perceptually-conceived information on the other hand. is not the consequence of a formalized process. Data are a factor, but then so are other aspects such as individual traits, culture, structure and political processes.

The relationship between data and information is, therefore, not so strong. This contradicts the objectives implied by the creation of management information systems and data processing. The former allows the individual access to established information, whilst the latter converts data into information. Since the production of information cannot be formalized there is no point in trying to achieve it. Systems should, therefore, be designed to produce the most accurate and readable data in order to enhance easy conversion into information.

When Ligomenides describes information as a fundamental force, he should be referring to data. Levitan, on the other hand, is correct in claiming information to be dynamic and continuously evolving. The logic of these two views can be seen when data and information are differentiated. It is only data which is time-independent, and therefore unchanging and static—a fundamental force. By this we mean that data retains a constant empirical value: a fact is a fact and cannot be altered. Information, on the other hand, can be altered and changed. It is time-dependent and thus its value can alter from one moment to the next. It is dynamic and evolving because it is determined by an individual's perception.

It is, indeed, time and utility which transform data into information. If at a particular time a set of data is useful to an individual then it can be described as information. Equally, at another time the same set of data may not have utility and therefore not be information. The data themselves have not changed in any way, remaining always constant. What has changed is the individual's perception of those data. Information is in reality the consequence of a complex psychological process which transforms perceived data, into usable thought inputs.

It is, therefore, data which the individual receives from the environment, the brain which transforms these data into information. No formal system can directly pre-empt this stage; thus no formal system can transform data into information.

Data can be transformed into different data, transmitted and received. For example, production data submitted to the cost department and processed into second generation costed production data, which are in turn submitted to the financial director. Information is transformed, in part, from data; it can only be received and never transmitted. The financial director may attach value to costed production data and thus receive it as information. If he/she passes it on, it will once again be as data, perhaps refined into third generation data. There cannot be information systems because information is so transient.

The individual's subjectivity, which can be developed by training and formal procedure, determines what is information. At another time that perception may be different. Information systems cannot work in this way. Their assumption must be that information is not transient but consistent.

Quite obviously information systems do exist in one form or another. They may not be what they claim to be but they are nevertheless working. Is it worth arguing the difference? The answer to that lies within ourselves. Our actions and behaviour, and consequently our organizations, are governed by our perceptions. If, therefore, we understand information to be permanent rather than transient then how we organize ourselves will be coloured by that. We will design systems which use information as a resource and expect it to be totally consistent.

PHYSICAL CONSEQUENCES

The difference between these two paradigms may seem subtle, and in terms of information usage it may well be. However, NBS research indicates that the effects upon organizational design can be major. Structural characteristics consequent upon individually-owned information, for instance, can adopt two extremes. On the one hand they can be organic and informal (adhocracies) whereby all members are seen as contributing equally, each possessing a particular range of techniques to interpret data, whilst at the other extreme there are formal and mechanistic organizations (bureaucracies). The rigidities of structure stifle any use of information as a resource, so by default such processes are left to the individual. These contrast with the resource-driven paradigm in which the structural characteristics of organizations are seen to be centrally orientated, with spheres of influence at the periphery, their importance dependent upon the perceived value of particular sets of information.

Larger use of Information

The development of an organization oriented toward the resource-driven concept of information is well known, and many large firms are aligned in that direction. A sophisticated network is established, more than likely computerized, accessing and distributing information. Particular attention is paid to ensuring standardization of information sets. For example, a piece of costing information is so devised that it can be used by accounts, marketing and production. These organizations treat all information as their property, attaching to it a value like any other asset. This denies the employee any sense of ownership and tends to detach them from the work process with consequential inefficiencies evolving. It is asserted, that a motivated staff is a contented and efficient staff.

One important way to motivate is to give the individual responsibility within their job. Such responsibility is manifest in how much ability a person will be given to determine the information need of their particular function. To standardize information in the way that many corporate systems do, takes away that fundamental element of responsibility. Levitan, on the other hand, would claim this to be an inevitable consequence of increasingly sophisticated corporate structures. In addition, Karen Levitan and others imply that information which is not used as a resource is inferior. On the contrary, one could argue that giving the individual ownership could result in a greater entrepreneurial drive.

Standardization is the killer of innovation. It is difficult to conceive of any substantial organization, however, where such a degree of individuality could be tolerated. Their size demands a need for cohesion through standardized practice. Therefore it is only in the smaller establishments where such freedom would work. Large organizations which do not treat their information as a resource still exist.

Their mechanisms of control, however, are not informal or organic. Indeed, the realities are the converse. Individuality is stifled by greater control, rather than less. No-one is trusted to use information for the organization's good without formal control procedures. Classic examples of such organizations are to be found within the civil service or large quasi-government institutions such as the post office.

It has been suggested that there is a relationship between the way in which a firm uses its information and its structure. A matrix can be established differentiating organizations in terms of how they treat their information and the consequential structural characteristics. If, for example, information is plotted against size (another important contingent of organization) four major structural types could be identified. Each is the outcome of the way a particular firm uses its information.

On the vertical axis of the matrix are the two extremes of information. On the horizontal axis are the two extremes of organizational size (Figure 2). Both information and size are considered to be contingents, that is variables which affect the characteristics of an organization. The four possible combinations of these two variables determine four organizational structures. Thus, each structure is an ideal for each of the four combinations. For example, a managerial structure is the appropriate structure for a large firm using information as a resource.

Resource	Managerial	Technical/ Professional
Treatment of information	Bureaucratic	Entrepreneurial
Perception	Large	Small

Structure parameters

Managerial	: larger in size with central/top control but with more departmental integration
Bureaucratic	: larger in size with formal structure and a diversified control
Technical/professional	: smaller in size with a professional body and formal structure
Entrepreneurial	: centrally controlled but with less formal structure and smaller in size

***Fig.3.5** : Information/structure matrix*

Research into the validity of such a matrix is by no means complete. There is evidence from the NBS regional survey, for example, that such a correlation does indeed exist. In general, the more successful managerial-style firms tend to treat their information as a resource, whilst the more successful entrepreneurial organizations tend to treat their information as a perceptual phenomenon, and so on. It is not yet certain whether the matrix is predictive, that is implying combinations which are not successful. There is sufficient evidence, however, to support the claim that information usage affects organizational structure. The issue of information is much more than whether or not it is a resource. Information is indeed the essence of an organization, shaping and determining its structure.

For example, bureaucratic organizations could not effectively treat information as a resource because their mechanisms could not control it that way. A technical/professional organization, on the other hand, is better able to handle its information as a resource because of its structural characteristics.

The implication is that there should be a natural balance between the design of an organization and its information structure. Where this has not been achieved it is because managers are not matching the structure of their organizations with an appropriate usage of information. NBS research has also shown that the implementation of computer technology can worsen any potential for imbalance. This is perhaps because of two factors.

Firstly, the characteristics of the technology itself do tend to enhance resource-driven rather than perception-driven information. Secondly, the greater volume of information generated makes the problem more apparent.

If the matrix in Figure 2 is valid then there is neither a right nor a wrong way to control information: much depends upon the individual circumstances. Resource-driven information and perception-driven information are opposite extremes of one continuum.

Management teams were grasping eagerly at the new computer technology as a solution to their efficiency problems. There were at that time forces in the environment—market, political, social or economic—with which the traditional methods of organization could not cope; it was believed that the computer would provide all the answers. In the space of that time many colleagues began to discover that the impact of computer technology seemed to be contradictory. On the one hand, it did tend to improve efficiency, whilst on the other it was not always so certain that an improvement in effectiveness followed. It seemed that underlying, and as yet not understood, forces were at work.

Departments employing the same technology were more or less successful than others. Business organizations with all the ingredients of success and implementing computers to further fuel their growth quite often experienced failure a few years later. Since those early days there has been considerable discussion and research on the impact of information technology. It seemed that the perceptual aspects of organization were not merely misunderstood but often ignored, and thus we were not able to understand fully the implications of introducing information technology. We now realize that not only does it have an impact on the way we run organizations, it also has an impact on our behaviour. There is a readily understood physical aspect of any firm and a more abstract, perceptual aspect which arises from the way that people behave as a collective. Information technology is not the only agent in this process, but because of its particular characteristics it can have an extraordinarily powerful effect on both these aspects. One of the two main objectives of this book, therefore, is to underline the importance of the relationship between, information technology (and to a lesser extent general technology) and its organizational structure.

The two are inseparable. Models that seek to understand either technology or organization must do so within the framework created by this relationship. Otherwise an essential factor leading to an appropriate analysis will be missed. The other main objective is to produce a model which can do this. Unlike many of its predecessors, the model will not concentrate on the functional or hierarchical nature of organizations but rather on the dichotomy between the physical and perceptual aspects. This exists whether or not there is information technology present, but in many cases the implementation of such technology can exacerbate potential dysfunctional forces existing between the two. In effect there are two organizations—a physical organization and a perceptual or virtual organization.

The impact that information technology can have on the sometimes tenuous relationship existing between the two can be quite profound. The model explores thoroughly this relationship. It proposes virtual organization as the antithesis of physical organization. That is abstract, unseeing and existing within the minds of those who form a particular organization. The point being that because the framework of virtual organization is often subjective and open to many different perceptual interpretations, it is difficult for anyone to establish an appropriate response to it. This makes it no less real; nor does it reduce the need to respond to it. Left unattended the effects of virtual organization can be quite catastrophic, especially if stimulated by the impact of information technology. This has considerable significance for the practitioner, especially those who have introduced, or are thinking of introducing, information technology. The model suggests that for a firm to survive there would ideally exist a dynamic equilibrium between the two organizations which has been termed the organizational balance. Anything which upsets this balance, such as the introduction of information technology, has to be compensated for by proactive action of management. If managers do not understand, nor indeed perceive the existence of virtual organization, then they will not be able to take the appropriate measures. Long-term structural weaknesses or even total failure of the business can result.

Demonstrating the existence of virtual organization can be problematic. By definition it cannot be observed directly since it exists only in the mind. At best one could ask a selection of staff for their impressions and make certain deductions, but these are likely to be dubious and tainted with the researcher's own perceptions. As such there has deliberately been no direct empirical research to establish the foundations of this model. Support for its validity has been sought in three other areas. Firstly, there is indirect empirical data gained from several research projects conducted at the Newcastle Business School. This comprises a study over two years by Stuart Maguire and myself into eighty-five small to large firms in all sectors of the North East. We examined particularly the strategy that firms employed to implement and control their information technology.

The data were gathered through a series of observations, interviews and involvement in the process. Other data were obtained from one or two firm studies on the impact of information technology by a group within the Newcastle Business School called AMBIS (Advanced Manufacturing and Business Information Systems). The argument for using these data rather than dedicated data is that they should be less contrived. Secondly, there is support to be gained from relevant personal experience. Although many would find this approach invalid because it does not comply with traditional or conventional ideas about research data, there is an increasing acceptance that such data can prove useful. In the social sciences, for example, Duke argued that such an approach (amongst others) would make a valuable contribution to the understanding of social experience. It is accepted that it cannot provide the only data, but the data that it does provide should be considered no less valid than any other.

We return to a more traditional style in support of the main arguments. That is, rather than gathering empirical evidence first and then developing a model, we should develop the model first from experiential instincts and then support it with data. This is put forward in the belief that to pursue the traditional methods of data-first research would not produce the generic model necessary to achieve the objectives set above. We have become unimaginative in our ideas about organization because we have been blinkered by too many sets of data.

How the book is structured in terms of presenting a logical argument takes on a greater significance than if it were based merely upon direct empirical data. It requires presentation of basic argument, counter-argument and a compromise position which leads the reader forward to the logical conclusion. This is the dialectic which involves thesis, antithesis and synthesis.

This overview, therefore, should not only cover the background to the book's major arguments and supportive material, but also provide the reader with a basic chapter plan.

To explain the impact of information technology we must look beyond the functional aspects of business organizations. The importance of information is reflected in its ability, through the appropriate technology, to corrupt departmental boundaries. This chapter, therefore, proposes that it is better to understand an organization as a network of information flows, which is itself set within other network flows of region, market, industry, and so on. Such a perspective is underlined by many firms accessing information beyond their established functional boundaries. Within the organization, this could involve different departments accessing a common pool of information from the company database.

Externally, employees can gain entrance to public databases such as Prestel, which are run by autonomous bodies. The drive is towards centralization and hence better management control. Paradoxically, the opposite is being encouraged, whereby every user has the potential to access information which enables them to operate more efficiently, and also to do so outside the control of their manager. Under these conditions it is difficult for managers to control not only who reads and writes what but also, and perhaps more importantly, who knows what.

Large amounts of information can be processed and communicated via a network without managers ever being able to determine whether it runs contrary to their own policy. Security has, as a consequence, become an extremely important issue. The more the dissemination of information is allowed, the more likely control mechanisms will become corrupted.

In simple terms, data are facts, whilst information is the consequence of interpretation. What is information to some is not to others, and it is the subjective interpretation of what is and is not valuable which has caused managements considerable problems. The control parameters on the organization's database are based upon managers' own perceptions of its value as information which may not coincide with that of a potential intruder or misuser. But more than this, individual employees' perception of their information stock widens because of their greater accessibility to the information network. As a consequence, their perception of their own job parameters also widens beyond what management might think desirable.

4

EDUCATIONAL TECHNOLOGY

We are yet to realize the goals set in our Constitution in the field of education. A lot needs to be done to meet the emerging needs of the nation. The Constitutional Directive to secure universalization of elementary education is yet to be implemented fully; democratisation of education is yet to be realized; equality of educational opportunities is yet to achieved and eradication of illiteracy is still an unfinished task. Although various endeavours have been made both at the national and state levels to realize those objectives and spectacular progress has been made during the last three and half decades, the country has still to go a long way in reaching the target.

With a view to re-orienting and reforming the educational system, a large number of suggestions have been made by different committees and commissions and the Government of India as well as the State governments have been trying to implement these recommendations. But these steps have been more sporadic and less systematic in the shape of patch-works only. Instead of overhauling the entire system, we are merely tinkering with the casual changes here and there according to expediency and exigency. This has resulted in unsatisfactory progress and impermanent solution.

Therefore, as suggested in the Report of the International Commission on the Development of Education, "Learning To Be", new educational strategies must proceed from an overall vision of educational system and must conceive of education as an enterprise transcending the framework of school and universities overflowing its constituent institutions. Flexible, integrated and systems approach has to be adopted for bringing about desirable changes in education.

The Philosophy

Educational system may be compared with an organism or a machine in which all the parts or components are interdependent and their interaction is dynamic. In an organism there are living cells which are acting in different directions individually and contributing towards common goal at the same time. In a machine, there are various parts like engine, wheels, steering, chassis, and so on which have their individual functions and at the same time helping the machine to do its job. Both the organism and machine have a particular purpose or goal to realize and all their parts are tuned or oriented or geared to achieve this goal. Thus, in spite of individual entities and functioning a common goal is aimed at coordination, cooperation and integration.

UNESCO has also given the example of human body in explaining a system. Besides a large number of living cells, different sub-systems and units like circulating system, nervous system, digestive system, seeing and hearing system function within the large system of the human body independently and jointly for existence of the body itself.

The system is thus "a complex", a set of "connected things" or "organized body of materials or things." It is an assemblage of objects or units with regular interactions and interdependence collectively contributing towards an overall complex function.

Systems approach or systems analysis is a systematic way of identifying goals of any system and scientifically working out different steps to move towards these goals. It has been aptly mentioned in the "Learning To Be" that "It is precisely a characteristic of systems analysis to integrate uncertainty into daily action. Be this as it may, to the extent that it enables us to orchestrate many agents into a unified process leading to the greatest possible efficiency, systems analysis would appear to be an intellectual instrument which may be applied to an overall critical study of existing educational systems and is likely to suggest new scientifically calculated pedagogic patterns."

The systems approach generally consists of the following steps:

(i) an analysis of the existing situation;

(ii) setting up goals for the desired situation;

(iii) defining mechanisms to evaluate the achievements of goals;

(iv) generating alternative solutions;

(v) choosing the best possible solution through cost-benefit analysis;

(vi) detailing out the design of the system;

(vii) outlining the monitoring mechanisms for the system;

(viii) working out for introducing the solution.

SYSTEM TECHNOLOGY

Systems Approach is an innovative and rational problem-solving method of analysing the educational system and making it effective. It views the educational process as a whole with its dimensions and components like pupils, teachers, curriculum teaching media and materials, instructional strategies, physical environment and evaluation of instructional objectives.

According to the systems approach in education, the entire teaching-learning process is taken as cybernetic or organic which is self-generating as well as self-controlling. As the general health of an individual depends upon working conditions of different subsystems like respiratory, digestive circulatory, *etc.*, so also the educational achievement and progress at a particular stage is proportional to the functional efficiency of various sub-systems like teacher, supervisor, administrator, librarian, evaluator, and so on. Integration and interaction among all these agencies are to be promoted for overall efficiency. The following steps are adopted in the systems approach to education:

(i) Definition of instructional goals, behavioural objectives and stating them in operational terms;

(ii) Determination of functions for achieving these objectives through the use of various media and materials;

(iii) Identification of learners needs and characteristics;

(iv) Selection of suitable methods for effective learning of the topic;

(v) Selection of appropriate learning experiences;

(vi) Choice of relevant materials, media and resources for such experiences;

(vii) Assignment of appropriate personal roles as teachers, students AV personnel, librarian, *etc.*;

(viii) Implementation of the programme starting with a try-out under typical conditions;

(ix) Evaluation of the learning outcomes in the light of instrument/ instructional and behavioural objectives; and

(x) Redefinition and revision of the goals, objectives, content methods, strategies, *etc.* for improving learning experiences.

With a view to improving the teaching-learning (input-output) process at the outset, the instructional objectives should be formulated and the operational behavioural objectives must decided in the light of instructional objectives. Adequate linkages should be established between the inputs and outputs and a feedback system be developed. This process is not merely cognitive. It operates in a social, emotional and affective environment. According to Panda, motivational and affective considerations have occupied a significant place in the learning process and outcomes.

INFLUENTIAL APPLICATION

The systems approach should be introduced as an integral part of the whole educational environment and accepted by all concerned with the environment. Otherwise, it will not be effective and its implementation cannot bring about improvement. Formative evaluation and benchmark survey may be necessary for formulating objectives, subject-matter and strategies, training and orientation of teachers and generating adequate awareness and imparting necessary knowledge as well as skills.

The teacher has to plan for utilization of resources, organizations of activities, use of media, *etc.* He should have a good knowledge of the subject limitations of the environment, individual differences of children and suitable methods as well as materials so that he can plan the system accordingly. The systems approach involves continuous evaluation, the results of which contribute to the revision of the plan wherever necessary.

The Benefits

Sampath and others have enumerated the following merits of system approach:

1. It provides a conceptual framework on which to build plans for implementing change for education.
2. It helps to identify the suitability or otherwise of the resources to achieve the specific goal.
3. It helps to assess the resources, needs, their sources and facilities in relation to quantities, time and other factors.
4. Technological advance could be used to provide integration.
5. It permits an orderly introduction of components demonstrated to be required for system's success in terms of students.
6. Rigidity in plan of action is avoided as continuous evaluation affords desired beneficial changes to be made.

Responsibility of Teachers

The teacher has to play a pivotal role in effectively adopting the systems approach not only in his teaching process, but also in organizing various projects, both curricular and co-curricular in the school situation. But there is ample truth in the old saying that it is difficult to teach the "old ones new tricks." It is also equally easy for the new teachers to forget the "tricks of the trade" learnt during the pre-service training period. Therefore, systems approach should be first in the educational planning and administration itself, so that no constraints and adverse situations are faced by the teachers. The teacher should act as a "friend, philosopher and guide" of students and implement all projects according to the principles of systems approach both in theory and practice.

Vocational Significance

The traditional style of education is burdensome and boring for the learners. Use of Educational Technology can make it interesting and learner-oriented. "The house of education which we want to build is of course going to be different than the cottage in which we are living till this time. We know the present cottage does not fulfil our needs at present and certainly not going to cater for the needs in the coming decade" said I.J. Patel, President, Indian Association for Programmed Learning in his keynote address at the Fourth Annual Conference on Programmed Learning and Educational Technology, Bombay, with reference to changing circumstances. What are these reforms then?

Educationists nowadays realize that in education 'learning' is more important than 'teaching.' Learning is concerned with pupils whereas teaching is concerned with pupils and teachers. In the olden days, a teacher was the only source of knowledge. The students learnt what the teacher taught. With the advent of textbooks and other learning aids, the teacher's personal knowledge though important, ceased to be the only or even the paramount source of learning. Hence, the textbook, author, the radio programmer, the film-producer and many others now assist in the learning process.

The traditional classroom with one teacher teaching 30 or 40 students which was mainly one-way communication of information, is no longer effective for modern times. The two general factors "information explosion" and "population explosion" are bringing about changes in the developed and developing countries. Both of them have posed critical problems for education. More things to be learnt and more individuals to learn. It is not possible to solve them by conventional means. For solving these problems successfully, educational technology consisting of various media of mass communication, suitable child-learning process and modern testing and evaluation techniques are required. Especially in developing countries like India, it has to be mastered and utilized by educationists, if they are to keep pace with each other and catch up with developed nations. As such, both quantitative expansion as well as qualitative improvement of education can be facilitated and accelerated with the help of educational technology. As Apater has succinctly pointed out, "Today a technology of education is being developed with the aim not only of making education more widely available, but also of improving the quality of education which is already available."

Growth of Syllabus

Any kind of educational improvement should be preceded by suitable curriculum reforms. Curriculum is the sum total of learning experiences provided by school. It is called the "stuff of education." It is what we teach, what the students learn, what we examine for and what we give degrees or diplomas for when students pass through. But we should make it relevant to and continuous with the changing needs of the present and challenges of the future. It must be defined in terms of what and how we want the students to know, do, think, feel and appreciate when they are in or outside the school. While determining objectives and defining their respective learning experiences, we must keep in mind what kinds of knowledge, skills, thoughts and feelings the children bring with them when they come to the school. We should also consider what kinds of things are happening to them outside the school or inside the society where they are likely to go and what they will probably do after their schooling.

Seth Spaulding (1970 p. 11) has very rightly observed that one of the reasons for current student unrest throughout the world is the fact that curriculum reform has not been continuous and the curriculum has not been entirely relevant, neither to what the student brings to it, nor to what the student expects from it. The student in today's secondary school and university is a very different kind of person from the student of a generation or two ago. The information explosion has enormously affected the younger generation. The secondary school student has probably learned more out of school than he has in school and the university student often feels much closer to the problems of his nation than does the professor who lectures to him. Yes the curriculum has changed little to reflect the changing nature of the students and the society in which education exists.

Which subjects constitute the curriculum for secondary schools? Many of the subject-matters may be outdated or irrelevant. But it must be agreed that we should teach our students to know, do, think and feel about democracy, population problems, pollution problems, and so on, for future development of the country.

On the basis of broad objectives, curriculum planning should be done and suitable subject content, behavioural outcomes and other learning experiences be put in for making curriculum comprehensive. It should be based on the findings of curriculum research and be enriched by interdisciplinary collaboration among experts on different subjects. All this would contribute to the design of a total school curriculum that is complementary as well as comprehensive. Spaulding has added, "This kind of curriculum building process would represent a scientific approach to the applied art of teaching. It would make possible a ' system approach' to curriculum planning (or at least a systematic approach) whereby various goals as seen by different sectors of society are considered and a matrix of goals agreed upon. These broad goals are then spelt out into behavioural goals in concrete terms. Suitable methods, strategies and materials would be suggested and tried out and then improved upon. Accordingly, teacher training programmes would be formulated and implemented for training teachers to enable them to manage with competence the new content, new media and new materials.

Aspects of Technology

Although in a narrow way educational technology means little more than the use of sophisticated hardware in teaching, including overhead projectors, tape-recorders, televised films, cassettes video-discs, gramophones, *etc.*, in a broad way it may be interpreted to mean the use of any new technique or methods of teaching. Technology of education would include the entire process of setting the goal, the continuous renewal of curriculum, the trying out and use of new methods and materials, the evaluation of the system as a whole, and the resetting of goals in view of the changing circumstances. It is also often identified with various devices and means or recording, storage, manipulation, retrieval, transmission and display of data and information with high efficiency and speed. They are considered as an integral part of a broad curriculum for research and development design amounting to tremendous potentialities in education.

Technological media and materials useful in the instructional process range from simple varieties for helping teachers to develop and present his lessons more effectively in traditional classrooms to sophisticated machines and mechanisms completely changing the classroom teaching structure and situations. A number of technological media and materials can be useful in both teaching and in the management of administrative data that is necessary in modern mass education. The computer, for example, can be used in the instructional process by efficient manipulation of instructional materials (computer-assisted instruction) and at the same time can be used to make new kinds of administrative strategies possible (making the complicated individual tests of students, the flexible scheduling of classes feasible, and so on).

Scenario of Technology

It goes without saying that the use of various new methods and devices is desirable for helping the teacher to do his work better whatever he has traditionally been doing for long time. With this aim in view the audio-visual movement has been launched for the past two generations. But due to lack of coordination and sincere efforts, the audio-visual materials have not been used adequately. Many of such aids have been found lying idle and dusty and some have been out of order for most of the time. Of course, there are some teachers who take keen interest in it and make optimum use of the audio-visual materials available with them. Spaulding (1970 p. 18) has rightly remarked, "These first generation educational technologies have suffered (and continue to suffer) from ineffective utilisation because they are usually not conceived of as part of an entirely new strategy of instruction which requires new kinds of administrative support, new kinds of curriculum materials (software) and suitable for presentation via the devices (hardware) and new ways of managing the materials so that they are easily available to the teacher."

In the beginning, the audio-visual materials were only put to marginal and individualized uses. Subsequently they were applied as a stop-gap treatment for deficiencies in the system. There was neither any coherent thinking nor a scientific organization of these materials in the educational process. The new techniques used were very slow in finding their way into schools. Their increased use has been mainly due to interest and initiative of certain teachers. Gramophone and radio have been used for singing and games; projectors have been used to illustrate history and geography, tape-recorders for oral expression, pronunciation and narrating stories either individually or in groups and television for teaching any subject-item. Thus educational technology materials have proved of help in clarifying concepts, stimulating group and individual activities, developing a collective critical awareness, changing attitudes, imposing a new structure or organization on certain subjects and encouraging originality and creativeness. Henri Dieuzeide (1970 p. 5) has also added, " The use of these techniques has even sometimes made it possible to progress beyond a mere change in the educational climate and, for example, to encourage problem-solving abilities either collective or individual or develop self-evaluation processes."

There has been no systematic use of technology in education. Pupils are exposed to "sporadic bursts of audio-visual information." Only the half-hearted attempts are made to apply the techniques to conventional school activities. Use of these aids,. as already said, depends entirely on the teacher himself and becomes meaningful and relevant only when carefully fitted into an educational pattern decided by him. Therefore, teachers have to be properly motivated and made interested in the use of such materials. And also mere interest will not help much. They have to be trained and oriented in the adequate use and maintenance of the materials.

Educational Technology has to be taken up as a comprehensive and continuous programme. It has, of course, to be used as a remedy for inherent deficiencies in the system, in one place to offset the teachers' lack of qualifications by regular broadcast demonstrations or by programmed documents for the pupils; in another to speed up the introduction of new subjects or new methods of teaching, to take over activities on which schools have fallen down, and so on. Teacher-training programmes can thus be improved by radio and television broadcasts.

In many modern countries like the USA, UK, and France, attempts were made through special programmes of radio and television to eliminate the maladjustment between the school and the community by making up the cultural handicaps of certain classes of children. These broadcasts are intended to prepare them for school environment and to make up the "cultural lag" of certain classes of people in the society.

It is often found that these devices or media have been hurriedly introduced and used without sufficient planning and forethought. With the plea of urgency they have often been put to use for which they were not designed. For instance, television has some times been used simply as a means of verbal communication without any visual content thus repeating the old-fashioned teaching practices like lecture method or verbal teaching learning by heart, encouraging passive attitudes, autocratic teaching, *etc.* Therefore, the media are criticized as "retrograde innovations" which have tended to displace or disguise problems rather than solve them.

Moreover, the half-hearted combination of technology with a traditional system, by stressing to the point of caricature the worst features of the system, has forcibly emphasized the need to re-examine its aims and its methods.

With the development of research in cognitive and child psychology, educational technology has come to be immensely influenced as well as developed. Modern psychology in general and Piaget's work in particular on the behaviour and aptitudes of children, learner's readiness and needs, has been felt everywhere. At the same time, Skinner's work on learning and the development of programmed learning have had an equally wide influence on education. The users of the new technologies have been led to state education problems in more precise terms as regards aims, organization of the subject to be taught, nature of learners' activities and evaluation methods in relation to educational aims and objectives. This Copernican revolution in teaching has transferred the centre of gravity of educational thinking and activities from the teachers' functions to the pupils' activities and as such has enlarged the prospects opened up by "Educational Technology."

Educational Technology can promote a certain amount of flexibility into the functioning of the school system which has been in a rut for decades. For imparting in-service training to teachers and improving their professional growth, educational technology provides ample opportunities. It also works as a means of rapid dissemination of education on a massive scale and of increasing the effectiveness of education by making learning a more individual process. Moreover, educational technology implies line methods for the production of teaching materials and a division of talent in their uses.

The development of educational technology would open up the prospect of creating and recreating new types of educational institutions in future radically different in structure and function from those of the traditional ones. It would help in reducing wastage to a minimum, which is a great drainage of resources both physical and human. Since the society of tomorrow is to be founded on lifelong education, the self-teaching or self-study centres would grow-up, which can possess various educational technologies. Even the formal teaching would no longer mean a matter of forcing information upon inert pupils in an authoritarian method, but one of exposing self-motivated youth to the practices of self-teaching or self-learning.

The misgiving that the role of teachers will be reduced, is not at all entertained by the exponents of educational technology. Rather it is affirmed that teachers will be made more receptive and will be placed in a more central position where pupils can more easily approach him with their individual problems. Henri Dieuzeide (1970 p. 13) an expert of international distinction on the matter has rightly asserted, "In this connection, we cannot stress too strongly the fact that the use of educational technology far from implying any qualitative decline in the role of the teacher frees him from certain purely mechanical tasks of exposition and repetition, thus enabling him to devote himself to the noble and irreplaceable functions of stimulation of interest, diagnosis, motivation and advice."

Education of tomorrow will thus be able to play its role more effectively by making the individual creative, innovative and efficient. Of course, success of education cannot be achieved merely by substituting mechanical methods for human beings, but by developing new patterns using both human and technological know how in order to teach more people better and more rapidly. On the whole, Educational Technology has the potential to bring about improvement in education both qualitatively and quantitatively.

Policy Matters

> "Educational Technology offers the means to reach numbers in remote and inaccessible areas, remove disparity in educational facilities available to the disadvantaged and provide individualized instruction to learners conveniently suited to their needs and pace of learning" according to POA (p. 183). All these technologies require adequate infrastructure supporting staff, trained personnel, suitable building (even in urban areas),

willing teachers, *etc.*, Since in the rural areas assured electricity supply is not available NPE, 1986 (p. 22) has suggested, "In villages without electricity, batteries or solar packs will be used to run the programme. Thus alternative arrangements need be made for avoiding frustrations and dislocation in the teaching-learning process."

The Ministry of Human Resource Development (MHRD) and the Ministry of Information & Broadcasting should jointly plan for long-term and effective use of media usage. The CIET and UGC should also make coordinated efforts in providing leadership and guidance in production of software, conducting research and evaluations as well as training the personnel for production and utilisation. The State governments and their agencies should take up production of locally relevant programmes and materials providing equipment and receiving systems, monitoring and evaluating the utilization of such technologies at various levels. Voluntary agencies of such technologies should be involved as far as possible in the production and utilization systems. Both hardware and software facilities should be provided by these agencies and mutual cooperation should be ensured for their effective maintenance and application.

During the 9th Plan the National Policy on Education, 1986 and POA, 1986 were reviewed and recast. Most of the SIETs were made autonomous. Particularly, the Central Institute of Educational Technology (CIET) and 6 SIETs in Uttar Pradesh, Bihar, Orissa, Gujarat, Maharashtra and Andhra Pradesh were strengthened and continued to produce ETV Programmes since 1988. During 8th Plan it was found that the working of SIETs and their output was not up to the mark. Hence, an attempt was made to improve the quality of production and increase their output. The ETV Programmes were mainly of an enrichment type. The number of Non-Government Organizations (NGOs) was not substantial so far in this field and an important initiative was taken in distribution of radio-cum-cassette players and colour TV sets in primary schools. Programme were also produced for Adult Education and Teacher Training, particularly, for orientation and enrichment. The lecturers in ET and Teacher Educators of DIETs were trained by CIETs and other Institutions. The teacher training institutions like DIETs, CTE and IASEs were provided with VCRs and colour TV sets and efforts were made to produce and make suitable software available for their optimal utilization.

At higher education level, IGNOU continued to provide a half-hour slot on the national TV network every day. Countrywide Class Room Programme of the UGC continued its transmission daily. The UGC created 15 Educational Media Research Centres (EMRCs) and Audio-Visual Research Centres (AVRCs) which produced programmes for higher education. The UGC also developed and set-up an Inter-University Consortium for Educational Communication (IUCEC) for providing coordination and leadership among the above institutions. Some NGOs also came forward to produce software material with the UGC assistance. In technical education, 4 Technical Teacher Training Institutes (TTTIs) and some IITs were provided reasonable production infrastructure and they produced programmes suitable for technical education. The Ministry of Human Resource Development made expert assessment of the needs of programme production and personnel training and made initial attempts for improving the existing conditions. With the availability of INSAT, more transmission capability was acquired and the GRAMSAT Project was implemented by the Department of Space.

Computer Education was promoted in schools under the CLASS Project. Computer Literacy Programme was encouraged in most of the schools by the professional agencies. The CBSE pioneered Computer Science Education at higher secondary stage and some State Boards followed suit. The UGC also started the programme by providing PCs, introducing diploma and degree courses in computer science and providing research and higher studies in this field. The CLASS Project expanded its area of operation. Besides MHRD, Ministry of Information and Broadcasting, Department of Space, Department of Electronics and Finance Ministry augmented transmission facilities for educational programmes and set up educational channels. The CIET and UGC co-ordinated and monitored training and production facilities and State Governments were involved in the funding and management of educational technology programmes. The CIET and IUCEs have involved various NGOs to give a fillip to the Educational Technology Programmes.

Role of Technology

Education Technology has become very important today in all the fields of Education. Its need was identified quite early by some. "Education, normally one of the largest employers in any country, is one of the industries which has not undergone a technological revolution. We must help to move it out of the handicraft stage. With the terrible and growing shortage of qualified teachers all over the developing world, we must find ways to make good teachers more productive. This will involve investment in textbooks, in audio-visual materials and above all the use of modern communication techniques (radio, television and film) for teaching purposes." This is remarked by Robert F. Mc Namara, former President of the World Bank addressing the Board of Governors in 1968.

What is the nature and scope of this "Technological revolution" to be brought about in the industry of education as visualized by the world famous economist? Why has he emphasized the use of modern communication techniques for teaching purposes? In what way will this contribute to community education or solve the problems of the community? These are the questions we have tried to answer in this chapter.

Grand Success of Technology

It is beyond doubt that technological revolution in education will occur only through educational technology which in a broader sense includes the entire process of curriculum development, the setting of objectives, the planning of curricular programmes and practices, the try out of new methods and materials, the evaluation and feedback of the system as an integrated whole and resetting of objectives on the basis of findings and experiences out of evaluation and innovations. Any kind of educational improvement should be preceded by suitable curriculum reforms. Curriculum is the sum total of learning experiences provided by the school. It is what we teach, what the pupils learn, what we examine for and what we award degrees or diplomas when students pass through. But we should make it relevant to and continuous in consonance with the current needs of the present and probably challenges of the future. It must be defined in terms of what and how we want the students to know, do, think, feel and appreciate when they are in or outside the school.

While determining the objectives and defining their respective learning experiences we should also keep in mind what kinds of knowledge, skills, thoughts and feelings the children bring with them when they come to the school. At the same time we cannot close our eyes to what is happening to them outside the school or inside the community where they are likely to go and work after their schooling.

Seth Spaulding (1970, p. 11) Director UNESCO Department of School and Higher Education has rightly observed that one of the reasons for curriculum reform has not been continuous and the curriculum has not been entirely relevant either to what the student brings to it or to what the students expect from it. The student in today's secondary school and university is a very different kind of person from the student of a generation or two ago. The "information explosion" has enormously affected the younger generation; the secondary school student probably learns more out of school than he does in the school and the university student often feels much closer to the problems of his nation than does the professor who lectures to him.

With a view to making the curriculum relevant and meaningful, not only necessary reforms have to be brought about, but also suitable media and methods are to be adopted without delay. This necessitates greater psychological and pedagogical preparedness, more social concern and community involvement, a new scientific attitude and a co-ordinated or integrated approach to the educational process as a whole. Here comes in the scope of Educational Technology which seeks to have a happy conglomeration of all the media, methods and materials used for better teaching and learning.

Thus Educational Technology comprizes methods from textbooks teaching to micro-teaching and deals with media and materials ranging from chalkboards to television or computer. As a matter of fact, technology of education involves books and blackboards, models and maps, tapes and slides, radio and televisions, projectors and computers. It includes both 'hardware' and 'software' and audio as well as visual materials.

It has rightly been pointed out by Geoffrey Hubbard (1974, p. 14) Director, National Council of Educational Technology, England "Educational Technology is the complement of curriculum reform concerned with the Method where curriculum reform is concerned with the content., Given an acceptance of the concept of educational technology as the process of improving learning process, we are faced with the problem, of encouraging innvoation in an educational system whose greatest strength is in decentralisation and autonomy."

This leads us to think that the methods of learning, *i.e.*, learning how to learn the habits of self-study, independent thinking, *etc.*, have to be emphasized. The learner has to develop a continuing concern for his freedom and his self-responsibility and his quest for self-fulfilment should come from within and without.

The International Commission on the Development of Education 1971 (UNESCO p. 75) has rightly observed, "Education is all the more democratic when it takes the form of a free search, a conquest, a creative art; instead of being as it so often is, something given or inculcated a present or a constraint." The method of independent learning or self-study is thus a direct corollary of the use of educational technology. Hence the technology of education is not only more scientific, but also more psychological and more pedagogical. With the advance of science and technology there are now several learning aids which have revolutionized the learning process, in particular and education as a whole.

Factors of Technology

Modern world is faced with two crucial problems:

(1) Knowledge Explosion and

(2) Population Explosion.

New Frontiers of knowledge are being opened every now and then and more knowledge is being added to the already accumulated mass of information through ages. Similarly population of the world particularly in the developing countries is increasing by leaps and bounds. These two problems have raised a large number of difficulties in the field of education.

For Quality of Education: It is estimated in the world as many as one third of the children of school going age have not yet been in the school. If at all they enter the schools, only less than one third of them will be able to complete primary education. It is also said that as many as fifty percent of the student population are only repeaters. Thus wastage and stagnation also lead to the increased number of illiterates and semi-literates. Many developing countries including India, in spite of their maximum investment in education, are finding it difficult to reach the quantitative targets like physical facilities, provision of teachers and enrolment of students.

Our country is a Sovereign Democratic Republic. Education is intended to prepare the youth of this country, for effective citizenship. Suitable elementary education, *i.e.*, education of children between 6 and 14 years of age is expected to make them intelligent and responsible citizens. That is why the fathers of our Constitution desired to make this education universal. But more than five decades are now over and we have not yet been able to achieve the target even for the age group 6-11 years. We are all aware of the reasons for this. The poor economic conditions, social superstitions and natural barriers like forests, hills, rivers, deserts, *etc.*, stand on the way to realize this Constitutional Directive.

Educational Technology like films, television and radio programmes will not only serve the students living in far flung areas but also make their education effective and interesting. This will help in imparting primary education besides increasing enrolment and reducing wastage and stagnation.

The standard of education is apparently going down day by day. Educationists, in a large number, are deeply concerned about the irrelevance and inadequacy of education in the modern society. Many portions in the curriculum are becoming outdated very rapidly; many contemporary issues are also not properly dealt with in the programmes and practices of schools. Unless the students are made thoroughly familiar with the burning problems of the society and sufficiently capable of tackling them successfully, they will fail to be true citizens of the world in future and their education will be deemed to be a failure.

In this context, educational technology will be of immense importance. Because successful teachers will be able to influence and motivate a larger number of pupils with the help of mass media, expertize of a limited number of persons can be made available to innumerable students; an inadequacy of physical facilities like laboratory expensive materials and many other things can be made up through the use of modern technologies. Consequently the quality of education will improve.

But we have not yet done this upto the mark. Seth Spaulding (1970, p. 10) has therefore said "Although education has shared in the creation of technological age, education itself has not learned how to use quality of its own institution. In attempts to 'catch up" quantitatively, education has usually attempted to multiply what has gone on in a pre technology system of schooling."

Thus the teaching-learning process will be improved to a great extent with the help of new media and materials through clarification of concepts, stimulation of groups and individual activities, development of critical awareness, changes of attitudes and encouragement of originality and creativeness. The role of educational technology, both for quantitative and qualitative improvement of education, has been very clearly stated by Henri Dieuzeide (1970, p. 15) Director of the Division of Educational Methods, Materials and Techniques UNESCO, "Educationalists in the developing countries will be more inclined to see educational technology as a means of rapid dissemination of urgently needed education on a massive scale to large groups, whereas the developed countries will see it as a means of increasing the effectiveness of education by making learning a more individual process."

EDUCATIONAL TECHNOLOGY

As discussed in the preceding pages, Educational Technology will not only revolutionize the entire educational process of the young learners but will also immensely influence life systems of the community. It goes without saying that young students of the present are the adult citizens of the future. Hence education of the young pupils is in a sense an integral part of the community education. It has been found that poor enrolment, irregular attendance, wastage and stagnation are largely due to mass illiteracy. If parents or guardians are literate and aware of the value of education, they will not allow their children to be illiterate or without education.

It is rightly said that "A literate mother will never allow her child to become a school drop-out. Therefore, in the early period of American Independence the popular slogan of the senators was "Let us educate our masters." This was accepted not merely as a slogan, but as a clarion call for the nation, not only in America but also in all countries of the world.

This problem can be solved with more systematic and well integrated schemes of non-formal education based on psychological principles, functional needs and interesting methods. In this context, educational technology with modern means, media and materials will be of enormous significance. The mass communication media like radio, TV and films have already been used in the rural community. The SITE and INSAT Projects have inter alia the fundamental objective of mass education and the TV has proved its potentiality as an agency of community education through various evening programmes besides morning programmes for children and UGC programmes for higher education.

In India the number of out-of-school youth in the age group 15-25 is very large. This is a group of vital importance for every country from cultural, demographic, economic, political and social points of view. It is therefore felt very necessary to develop a programme of non-formal education for this age group. The content of the programme will be psychologically sound and adequately need based. The various films radio and TV programmes suitably designed and produced for the purpose will make the non-formal education attractive as well as effective.

In order to make various developmental programmes like family planning, agriculture, health, *etc.*, quite successful, community education has to be given more emphasis. A study was made by the Community and Family Study Centre of the University of Chicago to find out the reasons for unsatisfactory results of family planning programmes. The overall conclusion of the study is the lack of effective communication of the message of these programmes.

The study therefore suggests ways and means to improve the use of radio, television, direct mail and posters. Yash Pal, the erstwhile Director of the Space Application Centre, ISRO similarly said about agriculture, "Millions of people learn better by seeing. The farmer is very conservative because he has done certain things for hundreds of years. He is not going to do things differently until he sees that these things are beneficial."

Factors of Orientations

Educational Technology has been a part of the wider mass communication media and materials. If we accept Marshal Mc Luhan's theories of technological determinism-namely that changes in the technology of communication lie at the bases of all social changes, then modern technologies of education will undoubtedly effect the community life and learning. Since our future society is going to be a "Learning Society" and education has been deemed to be a "lifelong process" we have to realize the importance of educational technology, for the welfare of the community.

Compulsion of Work

Education has been identified as one of the most efficient instruments of social change and an important means of bringing about national development. But the expenditure in education is yet to be treated as an investment by the planners and administrators at the helm of affairs. It is still to be realised that human resources are more valuable than the material resources. A host of economists like T. W. Schultz, John Vaizey, John Sheehan, H. G. Johnson have accepted the concept of human capital emphasizing thereby the expenditure on education as an investment.

According to them the procedure as followed for estimating physical capital formation can be followed for the formation of human capital, that is, by the expenditure to produce them. As the expenditures help in increasing knowledge and skill, they also increase the value productivity of human efforts and yield a positive rate of return. The important means of improving human efficiency are, *inter alia*, health facilities and services, on the job training, formal education and adult education programmes for workers.

A number of studies have also established a close correlation between the workers' education and their efficiency in terms of production. All the developed countries and even many developing ones have laid stress on workers education. For example, in Russia with a view to helping young specialists study systematically all the time. An extensive State-operated system of advanced training has been set up and a large number of refresher institutes, departments and courses are being organized throughout the year and in various places to keep them abreast of the advances in their respective fields. V. Yelutin, Minister for Higher Education and Specialized Secondary Education of the USSR has said that combining work and study, though not easy, is increasingly popular.

The government gives evening and correspondence students an annual 20-day to 40-day leave of absence from work with full pay, to do study assignment and take examinations, on a paid four-month leave of absence from work on the diploma thesis or project. Students also get leave of absence to take their final State examinations and the State pays part of their travel fare.

According to the International Commission set up by UNESCO in 1972, lifelong education means that business, industrial and agricultural firms will have extensive educational functions. The Commission has aptly observed, "Economic circles in industrialized countries have realised the need for improved training and generally agree that expenditure on it is a worthwhile investment. Many developing countries, however, still tend to under-estimate the importance of training-within-industry" (p.198).

They have further added that young workers with a professional or technical education still need complementary training, and also have to learn to adopt to the conditions and rhythm of work in industry. They can do so on the job. It would be logical for different firms to take responsibility for various types of subsequent training, such as specialization, improvement of qualification, courses preparatory to promotion, retraining and training of executive and other personnel in other higher skilled jobs. In so far as education is to be carefully planned to meet the needs of the economy, it is imperative for educational institutions and industries to work hand in hand. In many countries both private and public undertakings allocate a portion of their profits to educating and training their employees. The Commission has recommended that this practice should be more widely adopted.

Vocations and Education

Continuing education has been emphasized by the educationists for improving efficiency of workers. This means that individuals who are put on jobs after their general education may take up higher examinations as private candidates. Suitable arrangements of part-time evening and correspondence courses should be made for them. This will promote "educationalization of vocations" which implies that persons put on jobs first, subsequently are provided with facilities for higher studies. On the contrary, the present practice is "vocationalization of education" which connotes persons are given education in vocations and then are appointed on the jobs. Thus in near future "vocationalization of education" will be replaced by "educationalization of vocations."

Technical and vocational education should be provided largely through apprenticeship training and in-service training courses to be organized under the auspices of the industries and factories. Both government and private enterprises should impart such training on the job or in workshops. A number of industries or a complex of factories may in collaboration organize common training programmes providing for both theoretical instructions and practical skills. Based on the model of USSR, they may set up their own training institutions.

Education for Adults

In India there is a high percentage of illiterate workers. Unless they are provided with adequate facilities for getting themselves not only literate, but also educated, their capabilities cannot be improved and democracy cannot be a success. In order to enable them to effectively participate in the development programmes, it is imperative that they should be given suitable education on the job during their off-time. They need not go to educational centres, rather education should come to them. For many of them adult education is a substitute for the primary education, and a complement to elementary or professional education. For a few it may be prolongation of education for furthering their efficiency and capacity not only as workers but also as citizens.

As mentioned by the International Commission on Education (1972) the present facilities are lagging far behind the adult population's educational and cultural needs. Step by step progress is therefore not enough. What is required is a giant leap forward. This can only be achieved if both private and public bodies join in contributing their resources and efforts to the maximum. The National Adult Education Programme inaugurated on October 02, 1978 was therefore a promising step and utilization of both government and voluntary resources, intra and inter-departmental endeavours are aimed at achieving the target.

Technology of Education

Adult Education for workers requires specific strategy and methodology and social institutes and courses are desirable to bring about desired results. As an illustration, Yugoslavia's workers universities may be taken for the purpose where attempts are made to find answers to many questions arising in the field of workers' education today. These institutes provided both formal and non-formal education and open to all workers whatever their earlier training and education. Courses are offered at all levels ranging from elementary education to specialized training for highly qualified workers. There must be flexibility in courses which means that a worker may combine components of individual methods or group methods. The programmes may be adopted to the environment and diversified according to individual and socio-economic needs. Thus functional education for professional and civic growth is as important as the workers' general and cultural development.

As the workers spend major part of the days' time in the factory or office, they cannot be expected to be fairly receptive and interested in education. To change over from the world of work to the world of study is strenuous for them. But they possess immense wealth of experiences and are good conversationalists. Therefore they are to be motivated to discuss their experiences, problems and suggestions with the teacher freely and frankly. This will be the suitable method of teaching for these adults. The traditional ways of lecturing and teaching from textbooks will not be effective for them.

Besides, media and materials should be selectively used in the right place and time. All the means, methods and materials that may be used for learning effectively and efficiently are in general called educational technology. To make learning interesting as well as successful it is necessary to utilize suitable media and materials for adults in general and workers in particular. The International Commission for Education has aptly opined, "However, systematic use of educational technology has in some cases made it possible to step up work on earlier plans for educational development and so gain time.

It has also helped to redistribute and make better use of qualified teaching personnel, and to improve the system's internal efficiency and yield by reducing the number of drop-outs and repeaters" (1972, p. 211). Thus the educational technology is important in terms of money and time but also significant effectively in terms of success and efficiency.

Education and work are intimately connected and one influences the other to a great extent. Schooling is the main institutional experience shared by the young, while work is the principal institutional experience of adults. Most jobs and occupations have educational requirements for entry and advancement. In the industrialized societies work education has to be made effective for improving productivity. Exposure to work situations and experience in the actual conditions add to relevance and effectiveness of the work education particularly for the personnel in service.

Radio, TV and other media bring to their listeners and viewers programmes on current issues and problems and news on developmental activities. For women, for example, there are programmes on housecraft, childcare, nutrition, *etc.* For farmers there may be programmes on good agricultural practices, use of fertilizers, availability of new seeds, equipment, *etc.* Not only in developed countries but also in developing countries like China, radio and TV are being used in the field of workers education very effectively. For example, Television Colleges came into operation in some provinces and municipalities in China in the sixties.

As reported in the *International Encyclopaedia of Education* (Torsten Husen, 1989, p. 44) "In 1979, the Central Radio and Television University was set up, jointly sponsored by the Ministry of Education and the Central Broadcasting Administration. With 28 other Television Universities at provincial or municipal level, a national network of television higher education has been formed. These Television Colleges mainly enroll in—service workers, cadres, scientific and technical personnel, school teachers, army and also a certain number of senior high school graduates who have neither been admitted to formal universities nor employed."

In our country, regular broadcasts and telecasts are specially designed for the general orientation and continuing education of farmers and industrial workers. A project called "Radio School for Farmers" has been operating from a good number of response from the clientele.

With a view to providing school as well as tertiary or higher education to the workers of the firms and factories, housewives and otherwise employed persons, the concept of open, spare-time and own-time education came into prominence. The British Open University-model was adopted by the Govt. of India and the experiment began with the establishment of an Open University in Hyderabad, Andhra Pradesh in 1982 with courses leading to B.A.; B.Com., and B.Sc., degree Correspondence courses/lessons were supported by audio cassettes. Open University (IGNOU) was set up with open courses in various degrees and diplomas.

The Open School was established in New Delhi for providing secondary education and also primary education in some places for the deprived sections of the society. A scheme of non-formal primary education is being implemented in the rural areas for working children in the age group 9-14 years. The National Policy on Education, 1986 has also visualized the provision of Open and Continuing Education of all types. A variety of skill-based in-service education programmes are being implemented for the primary and secondary stage drop-outs.

In India the Central Board of Workers' Education with regional branches has been organizing education and training programmes and utilising various modern methods, media and materials for making their programmes more effective and interesting. Both "big as well as little media" in the words of Schramm, both projected and non-projected audio-visual aids and dynamic as well as activities based methods should be used as far as possible.

Modern media like radio and television would immensely influence the quality and quantity of workers education. The programme of Action, 1986 (p. 182) observed, "Large scale use of audio and video programmes in broadcast and non broadcast modes would also generate enormous demand for qualified manpower to work in educational media set-ups. In the long run, it would be desirable and probably essential that maintenance structures be decentralized, both organizationally and geographically, and local technician entrepreneurs, could be trained for such responsibilities."

Thus the extensive use of media would not only improve the quality of workers' education, but also create employment opportunities at different levels. Educational technology rightly acts as two-edge sword and exerts profound influence on education in general and o workers' education in particular.

All the audio-visual materials ranging from books to television, from chalkboards to computers have their own limitations. If they are not used in time and with due care, instead of serving as aids they turn to be obstructions and create psychological hazards. As literacy is necessary for understanding printed words, visual literacy is also essential for comprehending graphics, films and television. The workers should be oriented to enjoy and understand. Otherwise these adult learners will not be able to make any sense and the message that is intended to be communicated falls flat on most of the clientele.

Radio and television programmes may be specifically produced and used for the worker community. The printed materials, tape recording, slides and film strips are to be used for clarifying certain concepts and explaining the points through discussion and demonstration. Last but not the least in importance is the inter-personnel communication or face-to-face conversation for teaching workers. All these are not exclusive but complementary to each other and are to be used suitably as in an orchestra to have desirable impact on the adult learners. Adequate feedback and extension activities together make the educational technology effective, leading to imparting of efficient education to workers.

5

PROGRAMMED INSTRUCTION

Mankind has entered the information age from an industrial age that prevailed till a few years ago. This change is expected to have a profound effect on the way we live, make money and deal with wars. It would therefore be useful to examine the nature of information and the information technology that is at the root of future opportunities, challenges and competition. Information has been there since beginning. Information technology was also there. But it is computer, which has made information technology so popular.

Information Technology (IT) as the name suggests deals with the technology that is used to handle information. It concerns with the storage, communication and processing of information. The IT has become a key to the survival of the business houses. Computer is one of the major components in Information Technology. Its application has penetrated every sphere of existence of modern man: Airlines, railways, weather forecaste, medical diagnosis, entertainment, banking-the list is endless.

KNOWLEDGE AND DATA

We express values, properties and names of entities through data. For example Ram, age, 26, good, man, *etc.* These units of expression do not, by itself carry much meaning. They express raw facts.

The term "data" is the plural of "daturn", though "data" is commonly used for both singular and plural forms. Data can be defined as facts, observations, assumptions or occurrences. They are factual materials used as a basis of discussion, decision, calculation or measurement. Data can also take the form of numerical or alphabetical characters or special symbols (*i.e.*, +, =, /, *etc.*).

Information. Information is derived from data. We generally refer to data as the raw material which after being processed gives us information. The data by themselves lack meaning and are incomplete. For example, the data like Ram, age, 36, good, man do not signify any meaning if considered separately. But if it is said that Ram is a good man or Ram's age is 36, then the data collectively carry some meaning. Thus data must be processed and correlated to arrive at a form that can be interpreted by human beings. The interpreted data is information. However, information in turn can also serve as raw data in some future transformation. Hence, in practice, the terms information and data are used interchangeably.

Information is derived from a related collection of data. We can take up a collection of data about all people working in an office, consisting of their names, sex, date of joining, present salary drawn, *etc.* to get information.

Name	***Sex***	***Date of Joining***	***Salary***
Rahul	Male	01/01/65	16,500.00
Anil	Male	01/01/58	20,000.00
Rita	Female	01/01/70	10,000.00
Anita	Female	01/09/63	15,000.00

Data and information can be used to answer queries. The queries like how many people are drawing salary above certain amount and similar kinds. The table below illustrates the information derived by a query to list all the people who draw salary more than 10,000/-.

Name	***Sex***	***Date of Joining***	***Salary***
Rahul	Male	01/01/65	16,500.00
Anil	Male	01/01/58	20,000.00
Anita	Female	01/09/63	15,000.00

Information at some level might become data for another level. The daily production figures in an organization may be considered as raw data for the managing director to whom they have little value as such. If however, they are arranged for the month and the shortfalls computed against the plan, it becomes information that would prompt him to take some action so that the production rate is accelerated to match the plan.

Find out the data and information in the sentence "Spring is an enjoyable season". You can also try to write five sentences that bring out the information in the table given above. We can help you out in writing one sentence "Rahul joined the organization in 01/01/65 and gets a salary of rupees 16500."

Importance of Computer

Now let us discuss the role, a computer has to play with respect to some of the fields like business, education, entertainment, science and home, in more details.

Business: Computer have now permeated all aspects of business and the economy.

It is a sophisticated tool that can significantly increase productivity. In a manufacturing industry, certain steps are required in the development of a wide range of items. These steps are:

1. Preliminary designing
2. Advance design
3. Model development
4. Model testing
5. Final testing
6. Production and construction

In the past, preliminary sketches, design drawings and engineering drawings were usually prepared early in the design and development of new products and projects. When designers or engineers had a new thought, they would make some preliminary sketches to get the idea down on paper so that it could be analyzed more thoroughly. As the design was modified, additional drawing would be required. When the design was finally approved, further detailed production drawings were prepared. Thus, the preparation of drawing could occupy a substantial portion of the designer's time and effort. Computers now make possible quick changes and modification in the sketches made by an engineer. Interactive communication between designer and computer may continue until a design with a desirable set of characteristics is produced. Once the initial drawings are finished and displayed to the engineer's satisfaction, the computer may then be instructed to analyze the displayed design and report on certain characteristics. Such interaction between designer and machine is now relatively common. This method of designing is popularly known as Computer-aided design (CAD). Again if one imagined factory site even a few years back, a picture automatically used to zoom in thousands of people involved in jobs of huge labour and danger. But now the scenario has changed. Today robots controlled by computers are used to perform thousands of factory jobs that are dangerous or boring for humans. In fact, it has been found in a study, robots may reduce overall costs in manufacturing industries.

The impact of computers has been deeply felt in the financial sector also. One such sector is banking. Banks depend heavily on computers to process transactions around the world. Banks communicate with each other and send funds-transfer instructions through computers. Computers also help in dealings with millions of accounts. It helps these segments in their integrated communication with their own branches and that of peers. It helps in magnetic coding of cheques, in the use of tellers in some cases the tellers are being replaced by automated teller machines (ATMs). An ATM is a device that's located on or off the financial institution's premises to receive and dispense cash and to handle routine financial transactions 24 hours a day. The implementation of credit cards that are used for automatic and instantaneous banking transactions has become possible only for computers. The concept of "bank anywhere, anytime" depends on the computer. Recently impressed by the increasing number of customers owning personal computers, many banks like Bank of America, Citi bank, ABN Amro, *etc.* are now offering a special service: banking by personal computers. Customers who have computers can call the banks and use their computers to check on accounts; transfer funds from one account to another and even pay bills.

The advantages of using a computer are also available to small entrepreneurs. This includes office automation applications, interconnection of computers, e-mail, browsing, *etc.*

Science: Scientists are now able to use computers to conduct research into complex problem areas that could not otherwise be considered. Powerful computers control all the satellites, which are orbiting in space. Even the instruments in satellites that perform experiments and transmit the results to earth are operated by computers. The data sent by the satellites are received from worldwide tracking stations, are processed and the information is recorded. This information, in turn, adds to the storehouse of scientific knowledge, leads to plans for additional space experiments and serves to reinforce or change existing scientific theories. Computers are also used for environmental control purposes. The National Centre for Health together with the Geological Survey is using computers to evaluate and control the level of pollution. To cite one example from the many, the Empire State System of New York collects water and air data from monitoring stations located at critical sites around the state. Data from the stations are automatically forwarded to a central computer. Each air-monitoring station reports every 15 minutes, and each water station transmits once every hour. Upon receiving transmitted data, the computer edits the message, sends any necessary operating instruction to the station, compares edited information to acceptable environment standards, and, if standards are not met, sends an appropriate alarm message to the Department of Environmental Conservation. Computers are also helping to conserve natural resources. California's State Water project, for example, is designed to conserve water by moving it from surplus areas in Northern California to needy areas in the south and west. All these facilities are monitored by computers located at five remote control centres. Many tasks such as weather forecasting were at one time considered impossible because of the time span and efforts involved in performing the necessary calculations. Now with the application of computers, such jobs have become possible. Another example could be the application of computers in astronomical calculations and computations involved in the modelling of chemical reactions for producing new drugs.

Medicine: Computers are now used for planning and control purposes by health care professionals. Applications of computers that have medical planning and decision making implications include computer-assisted diagnosis and research, and medical history preparation and retrieval. Doctors are using the computer as a diagnostic tool in hospitals and clinics. Nowadays in a diagnostic centre, only some physical examination take place, which consists of some basic tests. Data from these tests are fed into a computer. Within a few minutes after the examination procedures are completed, the computer output is ready. The computer may also be programmed to suggest tentative diagnosis to explain abnormal test results. The patient's physician, of course, is responsible for the final diagnosis.

Computers are also being used for such diagnostic purposes as displaying body functions on a terminal screen from motion-picture x-rays and other technologies and calculating the volume and width of the patient's internal organs by means of a Computer-Aided Tomography (CAT) scanner, Magnetic Resonance Image Scanner or Positron Emission Tomography scanner. These devices take pictures of body structures and then combine the thousands of shots into a single picture of the area in the body that has been malfunctioning, so that proper medication can be given right away.

In addition to being a diagnostic tool, the computer is also a research tool that's providing insights into the causes and prevention of stroke, as well as causes of drug addiction.

In developing a record of a patient's medical history, a usual practice is for the doctor or nurse to ask the patient a series of questions about past illnesses or health problems. This history taking is a time consuming aspect of the patient-physician relationship. Computer can be used to reduce the time involved and to tailor the questions to the patient's situation with the help of the expert system.

Education: Computers are becoming increasingly useful tools of education. It can bring to the educational process such attributes as untiring patience, around-the-clock availability. For some applications, such as routine instructions, computers can be superb adjuncts to the teacher. The computers are also being used for educational simulation in various subjects. For example, a programme can be run which simulates an Apollo moon landing. This simulation programme would help a student to implement what has been taught in the theory class. Not only this, the computers are ideally suited for delivering self-paced instruction where the course objective and contents are well defined.

The learning process of many subjects can be enriched because of the scale and range of information provided by computer data banks. New methods of training like Computer-assisted Instruction (CAI), Computer-based Training (CBT), *etc.* have since become realities. Let us discuss these new methods of teaching in more details.

CAI refers to a learning situation in which the students interact with and are guided by a computer, through a course of study aimed at achieving certain instructional goals. In a typical CAI setting, the student sits with a personal computer and communicates with the programme. Interaction may take place in the following ways:

- The computer presents instructional information and questions.
- The student studies the information or instructions presented and answers the questions.
- The computer then accepts, analyzes and provides immediate feedback to the student's response. It also maintains records of the student's performance for evaluation purposes.

The simplest and most-used form of CAI is the drill-and-practice approach that's designed to complement instruction received from teachers and printed materials. Student response is given to factual questions presented by the computer. Learning is facilitated because the computer can quickly supply correct answers as feedback to a student's mistakes. The drill-and-practice approach has been found useful in learning areas such as mathematics, statistics, languages, reading, spelling, *etc.* where substantial memory work is required.

A second and more complex level of interaction between a student and a CAI computer programme is found in the tutorial approach. With this approach, the programme assists in presenting new material to the student. The intent of the tutorial approach, of course, is to have the computer programme judge the actions of a private and very patient tutor working with an individual student. Numerous sub-options in the CAI programme permit a choice of materials to be presented, depending on how a student responded to the previous question.

Another method of education that is gaining popularity is Computer-based Training using multimedia. To put simply, multimedia Computer-based Training (CBT) is the use of computers and multimedia technology for training in a way that promotes student interest and motivation. The computer's multimedia capability to show graphics, run simulations, and play sound, video and animation greatly enhances the learning experience. Multimedia creates a complete multi-sensory learning programme, allowing students to interact with the material, and to learn according to their own needs, pace, and learning styles. Rather than replacing instructor-led training, CBT's flexibility and economy augment traditional training methods. In fact, the very nature of computer-based training—its availability at any time and any place—makes it an ideal resource for on-the-job training. The delivery of training to remote locations, multiple offices, and to a variety of training levels are all needs that CBT can satisfy on both performance and budgetary levels. Computer-based training can be delivered on a floppy disk, or over the internet.

There are many advantages of using computers in educational institutions. They provide one-to-one interaction with a student, as well as an instantaneous response to the answers, and allow students to proceed at their own pace. Computers are particularly useful in subjects that require drill, freeing teacher's time from classroom tasks so that a teacher can devote more time on developing new materials and doing research. A computer programme can be used diagnostically, and once a student's problem has been identified, it can then focus on the problem area. Finally, because of the privacy and individual attention afforded by a computer, some students are relieved of the embarrassment of giving an incorrect answer publicly or of going more slowly through lessons other than classmates.

Home: Computers have also entered homes in large numbers. Most of the home computers are used mainly for entertainment, but they have great potential for education, household management and household control. Today, the computer is fast becoming a household gadget, which combines a TV, a telephone, an answering machine, a fax, a CD player and internet connection, all in one. Apart from it, the computer offers a lot of entertainment for the entire family, like games. A key feature in the entertainment and educational aspect of personal computer is interactivity; the user is a participant rather than a passive observer. Remote working or tele-computing, *i.e.*, doing a job from home is also on the rise. There is also a trend of "small" appliances of various types. Besides microwave ovens and fully automatic washing machines, there are programmable recorders and radios, home security system and intelligent telephones, all controlled by built-in computing devices.

Entertainment: Another area computers have had an impact on is the entertainment area. Many people like science fiction or action movies, especially the one with the special effects. Well, those graphics are mostly made by computers. Most of the movies today use computer graphics to make things more realistic. Computer graphics and special effects are mostly used on games like spaceships, aliens, monsters. In the film Godzilla, the monster was created by computer animation, texture and graphics to make it more realistic than the older version where they used a man in a costume to play Godzilla. This wasn't the only movie made with computers. Movies like Jurassic Park, Independence Day, Star Wars, and the Star Trek used computers to make them look more interesting and gorgeous.

There are even movies completely made by computers like Toy Story and A Bug's Life. Not only movies used computer animations and graphics. Video Games nowadays use computer graphics and animations to make the games interesting and challenging. Computer games like Wing Commander Prophecy, Shogo and other new 31 games were made by computer graphics. It would be really boring without all this. That's how computers have had an impact on today's entertainment.

In the following section we take up a case study of office automation in details from a large pool of application of computers.

Progress in Automation

Different office activities can be broadly grouped into the following types of operations.

Document Capture: All documents originating from outside source like incoming mails, notes, handouts, charts, graphs, *etc.* need to be preserved.

Document Creation: This consists of preparation of documents, dictation, editing of texts, *etc.* and takes up major part of the secretary's time. Even a small mistake in an important document results in retyping of the whole document.

Receipts and Distribution: This basically includes distribution of correspondence to designated recipients.

Filing, Search, Retrieval and Follow up: This is related to filing, indexing searching of documents, which takes up significant time.

Calculations: These include the usual calculation functions like routine arithmetic operations for bill passing, interest calculations, working out the percentages and the like.

Recording Utilization of Resources: This includes, where necessary, record keeping in respect to specific resources utilized by office personnel. All the activities mentioned have been made very simple and effective by the use of computers. The application of computers to handle the office activities is also termed as office automation.

Well, the missing link is the calculation and recording utilization of resources. The conventional equipment like typewriter, has already become obsolete. Computer hardware and software have helped in building the following office automation devices.

For Information Capture and Copy

- Word processor
- Intelligent copy machines
- Document readers
- Digital voice systems
- Phototypesetters

For Information Storage

- Electronic filing system
- Micrographics

For Communication

- Computerized phone system
- Electronic mail

- Facsimile transmission
- Teleconferencing facility

Needless to say, every year new devices are getting introduced to make the office environment more efficient.

The principal features of the above mentioned devices are described below:

Word Processing + DTP: Word processing is often the "backbone" application for office automation. It is used to create, view, edit, manipulate, store, retrieve and print text materials.

The spectrum of word processing functions can be divided into two broad categories, namely;

- Text editing
- Print formatting

Text editing is basically the ability to enter, edit and delete textual matter. As text is keyed into the computer, it is displayed on screen. Sentences, paragraphs and larger blocks of text can be added, deleted, or moved around in the document being prepared.

After the text is correctly typed in, it is the job of print formatting. The Word Processor can be instructed to put heading and numbers on each page of a report, centre headings on the page, and produce specified top, bottom, left and right margins. The text can also be saved on the diskette for future use. When reformatting becomes necessary, it can be done by simple keystrokes. Remaining adjustments take place automatically. For example, you have a document which is made for standard page size. For the second copy, if you want to reduce or increase the paper size, all you have to do is, with simple clicks, reduce the paper size on the monitor as desired. Rest of the changes like the adjustment of text to fit into more number of pages, numbering of pages, placement of heading on newly created pages, takes place automatically.

In addition, several pre-recorded paragraphs can be called up from the dozens of files that are stored and can be assembled to create a new document. A "dictionary" or "spelling checker" programme can be called up that will check every word in the document against a list of correctly spelled words. If the word is misspelled, it can be corrected. This, global search capability can also be used to search for words or phrases specified by the individual. When found, they can be deleted or replaced with other words or phrases.

Try filling the gap in the next sentence:

The field of word processing function involves print formatting and ———. You must be remembering, it is the text formatting function.

The features that save you from hassles are tools available for spell checks, grammar check, thesaurus, automatic formatting of text and graphics, and many others.

Following are some more equipment/methodologies for computer usage in office automation.

Setting up of Files

Filing and archiving are often long, tedious, unpleasant and potentially dusty procedures.

Moreover, they are linked to considerable financial risk in the case of lost or misfiled documents. Computers are potentially capable of storing, in a safe and rapidly accessible manner, a digitised document. It is therefore natural that since the introduction of computers in offices, there has been continuous discussion of the paperless office. Another objective of electronic filing system is to save on costly office space occupied from file storage. The first operational electronic filing systems were put into operation in the early 80's. Two types of systems are currently available these days. One is based on a VHS cassette, and the other on an optical disk.

In the VHS cassette type system, a scanner copies the document and displays it on an A4 screen. Thereafter, the document is stored on a VHS cassette from where it can be called for reproduction by means of laser printer when required. One cassette can store upto 1000 A4 documents. Unauthorized access is guarded against, by multi-level passwords.

The optical filing system is used when large number of documents are to be filed. In this system; the scanned document is simultaneously displayed on a screen and written in a digitized form on an optical disk. Storing of an A4 document takes about 3 seconds while retrieval takes about 5 seconds.

Copying by Machine

These can be viewed as an advanced form of normal photocopiers that you normally come across. These can automatically load the copier, sort copies, and flip them over for making copies on both sides of the paper. Some of these can make multicolour copies, make borders, lines and logos.

Reading the Document

Using these devices, the existing paper document can be converted into electronically stored data without having to retype characters. The document readers can thus help remove large backlogs of otherwise inaccessible manual documents into an automated system.

Setup of Video

This system accepts messages by voice and can retain the same on magnetic media for retrieval, when 'needed. At the time of retrieval, the messages can be played back in exactly the same manner as they were recorded.

System for Photographs

Under most cases you would not want a computer output to look like a computer output. Phototypesetting allows high quality (book like) printing of your documents stored in a computer.

Use of Micrographics

You can use computer controlled microfilming devices for storage of legal documents, drawings and catalogues. The images can be stored on videodisks for fast retrieval, easy changes and secure access.

Computer in Communication

The computer with its capabilities of speed and accuracy can serve as an efficient user and controller of telephone lines. These systems, which include PBX equipment can handle voice as well as digital data in a highly efficient manner. They can provide electronic dialing, setting up of conference calls, add additional parties when the call is in progress. In case the call is not answered within specified time, the call can be routed through an alternate path. This system can also be used to maintain statistics about the phone calls.

Correspondence and Electronic: This technique can be used to remove inter office paper messages/memos. The technique involves sending the messages in a digital form, using telephone lines or dedicated lines. The message can be received almost instantly, thus avoiding the unnecessary delay caused by human involvement.

USE OF FAX MACHINES

Facsimile or fax refers to the transmission of text or fixed images and drawings by wire or radio channels. Fax permits the transmission of a mix of documents-including the handwriting, graphs, pictures and maps-that often cannot be transmitted at all over other communication media. It is very similar to Electronic Mail, except for that in this case, the message is sent exactly in the same manner in its original form. It appears like a photocopy of the original at the receiver's end. The transfer is immediate. There should be a fax machine connected to a telephone at the receiver's end too. The process involves placing the document to be transferred at an appropriate place on the fax machine, dialing the telephone number to which the destination fax machine is connected. Once the other fax machine responds (you will be able to hear the beeps, once the fax machine takes over the telephone line), you simply have to press a button and the transmission begins. Once the entire document has been photocopied at the other end, you will receive an acknowledgement at your end indicating that the transmission was successfully carried out. The whole process now a days is done through a computer.

In addition to fax transmission, there exists two other forms of electronic still-images transmission that have been adopted in several parts of the world. These two forms of still-image transmission—teletext and videotext are usually employed to access information from computer databases. In teletext, still images are transmitted in several scan lines of a television signal, which may be sent by either radio or cable. In the teletext receive, the still images are captured line by line and stored in local memory for subsequent display on a television screen.

In videotext, still images are transmitted digitally over the public switched telephone network by use of a modem. A modem is a modulation-demodulation device that converts the discrete stream of digital "on-off' electrical pulses used by computer, into continuous analog wave patterns. At the receiver's end, the digital signal is recovered from a modem and is stored in local memory, again for subsequent display on a television.

Benefit of Teleconferencing

A teleconference is a telephone or video-meeting between participants in two or more locations. It is a facility where using audio-visual equipment like cameras, microphones, *etc.* together with communication facilities, you can hold meetings of various people sitting geographically apart. All the members of the meeting will be able to see and hear each other, with the help of cameras and microphones, sitting at their own offices. This can vastly improve office productivity, especially as the commuting time and money to the meeting venue can be saved.

Most of these devices can be integrated into multi-function work stations, vastly improving office productivity. Indeed, apart from clerical staff and secretaries, such workstations, when provided to managers at their desk, can afford direct access to data.

Dark Side of Computer

Unemployment: It is feared in many quarters that computer is gradually replacing humans in the field of work. The greater efficiency with which a computer performs has resulted in job obsolescence. But it would be totally wrong to conclude that computer is the cause for unemployment. In reality, computers have opened up new job opportunities like programmers, system administrator, data entry operator, marketing personnel, Multimedia developers, web technologists, administrators, media experts, *etc.*

Trend of Depersonalization: In most computer-based systems, the record key used to identify a person is a number, *e.g.*, a social security, student, employee, or credit customer number. As people have come into contact with more computer systems, they have been identified with more numerical codes. Although many understand that being treated as a number results in efficient computer processing, they would prefer that systems be designed in such a way that they are treated as persons rather than numbers.

The Fundamentals

Attempting to incorporate a significant amount of computing in any coursework in the School of Education at this time requires the use of batch processing on the mainframe system because of the limited availability of terminals. The only resident equipment within the School of Education was an ancient keypunch machine. The computing facility itself is approximately two miles distant. The first attempt to break out of the mainframe "mindset" began with a request for resident equipment for the school of education.

When looking up a synonym for the word "web," one is likely to run into the word mesh, which then leads to the word netting which in turn leads to words like concealment, camouflage, cover, and blind. "Blind" might accurately capture the way many people feel as they wander through the World Wide Web and surf the Internet. This wonderful, yet sometimes perplexing, new technology has a variety of properties such as nonlinearity induced by hyperlink connections that can lead individuals on an indirect path that branches out exponentially rather than the traditional straight and narrow road. Unfortunately, this maze-like trek through cyberspace has the potential to cause anxiety in numerous individuals. The general area of computer anxiety has been well documented and studied; however, the more specific form of anxiety, Internet anxiety, is still too new to have been fully explored and researched. The purpose of this study was to focus specifically on Internet anxiety and determine what instructional techniques and behaviors either reduced or exacerbated anxiety in an Internet class for novice adult students. The Internet, which allows people to communicate through their computers, no matter the location, has become and still is becoming a global phenomenon of impressive proportions. One of the most well known aspects of the Net is the World Wide Web which is defined as "a convergence of computational concept for presenting and linking information dispersed across the Internet in an easily accessible way". This linking is responsible for the non-sequential properties of the system that affords people a high degree of choice when traveling through it.

From its limited and covert beginnings as a cold war communications system, the Internet has now grown to serve over twenty-one million households. It is thought that the numbers will grow as new technologies come into play such as a digital system, offering a new level of access, and Netscape, providing a new level of security which should entice individuals and business to conduct more and more transactions on the Net. Thus, the Internet is a technology that is becoming indelibly interwoven into society and unfortunately, one that will antiquate those individuals who do not learn to maneuver in cyberspace.

Although the cognitive aspects of Internet learning have been researched, the affective aspects of Internet learning such as feelings of anxiety have not been studied until know. Thus, the following definition and review of the literature will come from the body of work relating to the broader category of computer anxiety.

Knowledge of Computer

Computer anxiety is a fear of interacting with computers that is disproportionate to the actual danger of the situation. Computer anxiety leaves the user in an uncomfortable mental state in which he/she experiences debilitating physical and/or emotional symptoms. The type of fear experienced by the computer user varies from one individual to another. For instance, a person might fear breaking the computer, loosing important data or experiencing embarrassment in situations that he/she feels might arise from lack of computer mastery. After failing to adequately cope in an anxiety-provoking situation, the computer user may experience learned helplessness, a feeling of incompetence, inadequacy, and powerlessness. Learned helplessness can become a problem, since individuals who feel helpless will tend to avoid the anxiety inducer. Besides anxiety's relationship to avoidance behavior, computer anxiety is also related to poor performance. Anderson found that college students who failed a computer test had significantly higher levels of computer anxiety. Causality cannot be determined in this case, yet whether anxiety played a part in causing students to fail the test or failing the test increased students' computer anxiety, those students reporting anxiety cannot be allowed to slip away and drop out of the computer age.

The Fundamentals: The area of study known as computer anxiety, in several cases, including Todman and Monaghan's work has links to Albert Bandura's social cognitive theory, particularly with regard to students' self-efficacy. Self-efficacy is defined as a person's judgment about his or her ability to perform a task within a specific domain. The tie between low self-efficacy and anxiety makes Bandura's concept significant to the study of computer and Internet anxiety. Students, with low self-efficacy are more likely to: attribute their failures to low ability; feel they are less in control of their environment; and avoid the activity in the future. Given the attributes of low self-efficacy, it comes as no surprise that anxiety is often inherent to it. Students who judge their ability to perform a task as being low tend to "experience stress, anxiety and depression when goals have not been met". Thus, when studying students' computer or Internet anxiety, it is important to look at instructional techniques and behaviors that either lower or raise students' self-efficacy so that instructors can learn what to avoid and focus on effective strategies to employ.

As the Internet has grown so has the need for effective approaches toward curriculum and instruction pertaining to the Internet. Often articles present models for teaching and effective instructional techniques; however, they do not so often address Internet anxiety specifically.

The following are some instructional approaches that have been used in Internet classes: teaching students about both the hardware and software components of the system and information skills simultaneously providing hands-on experience; giving one-on-one assistance; attempting to make the students' first experience a successful one; asking students what they expect out of the class; encouraging students to share "questions, problems and triumphs" with the whole class; using step by step work-sheets as a supplement to verbal lectures; and encouraging peers to help one another.

Several researchers have proposed instructional strategies that are intended to take the "byte" out of computer anxiety. For some of these strategies, there is a high degree of agreement among professionals, while the verdict on other strategies remains elusive. The following are just a few examples of some of the most talked about ways to reduce anxiety or stop it before it ever starts.

An antithetical relationship exists between playfulness and computer anxiety. It is thought that play might be of particular benefit for people who are in, what is called the inquiry frame, which is the self-directed stage; people enter when they become focused on expanding and deepening their computing-knowledge. Others however, have found play to be beneficial in the beginning stages of learning. Pina and Harris, who focused on anxiety-reducing and confidence building teaching strategies to be used in computer literacy courses with pre-serviced teachers, cited play, such as allowing students to press any button on the computer as one of their main strategies.

Learning through verbal lectures and learning by doing are certainly not synonymous experiences and should not be misconstrued as such by instructors. For instance, a student listening to a lecture about a process on the computer may not be able to translate those words into actions which could cause a fair amount of anxiety for the student hence, the need for hands; on practice. Several researchers studying computer anxiety have documented the benefits of hands-on experience. Castleman and Pina and Harris promoted the hands-on approach in a non-threatening environment as well as Hunt and Bohlin, who also advocated providing demonstrations and ample hands-on time under the supervision of "tirelessly patient" instructors so that students can sharpen their skills.

Much of the literature in the area of computer anxiety focuses on unpressured instruction that increases self-efficacy and beliefs about the controllability of the situation. The notion of self-efficacy stems from Bandura's social cognitive theory, while the idea of locus of control comes out of attributional theory. Self-efficacy is defined as an individual's judgment of his or her ability to perform a specific task. Thus, a person with high computer self-efficacy would have confidence in his/her ability to master it. Locus of control, on the other hand, refers to a person's belief about the controllability of a situation. Those who view a situation as out of their control are frequently subject to anxiety and avoidance whereas those who believe they are in control tend to be more persistent and put forth more effort. In a qualitative study, Todman and Monaghan concluded that regardless of age, a relaxed, unpressured introduction to computers, in which the individual is encouraged to feel competent and in control, tends to result in a lower level of computer anxiety and a greater tendency to use computers subsequently as a student.

They define a relaxed environment as one in which stringent requirements and evaluation methods are avoided and methods to improve self-efficacy and controllability beliefs are sought.

The verdict is still not in with regard to self-regulated learning as a method for reducing computer anxiety. Pina and Harris advocated self-regulated learning and said it met with much success in their classrooms. However, in a different study that ran counter to the authors' expectations. It was found that a group of university students, who received computer training through direct teaching, exhibited less anxiety at the conclusion of the study than a group who received training through a combination of direct teaching and collaborative self-learning. The direct instructional model was considered a traditional model that included the presentation of new material, practice and review; where as self-regulated learning was thought to be consistent with a constructivist philosophy and a social cognitive perspective that emphasized students' employing appropriate strategies to achieve challenging goals which they have set for themselves thereby increasing self-efficacy. The researchers, in this case, utilized a quasi-experimental design together with qualitative methods such as interviews, adding depth to the study. Because of the relatively small sample size (group 1, $n = 16$, group 2, $n = 15$), caution should be exercised with regard to the generalisability of these results. Thus, it seems that more research needs to be done in this area to clarify the outcome.

It has been shown that because of the widespread and still growing use of the Internet, it is necessary to provide an Internet education to as many people as possible, in the best way possible, so that they will have the motivation to seek new computer knowledge and experiences in future. Part of providing this education in the best way possible is to acknowledge and attempt to alleviate computer anxiety in the classroom. Several of the instructional strategies used to deal with computer anxiety that have come out of the research, were discussed in this review. Although many of the strategies that are meant to reduce computer anxiety may be applicable to Internet anxiety, there is still a point where the two types of anxiety diverge. The one-of-a-kind virtual world of the Internet is a place without boundaries and a space without a core that will cause users to experience different kinds of emotions, some of which will be more negative than positive. For this reason, it will be important in future to address a more specific form of computer-anxiety-Internet anxiety.

In order to provide a vivid description of a particular site regarding a nascent concept-Internet anxiety—the decision to use qualitative research methods, supported by observational data, interviews and documents, was made. Qualitative research or field research beneficial are often used for detailed analysis of small group situations. In order to maintain confidentiality, all names presented in this study have been changed.

The subjects in this study were adult students in an Internet class designed for beginners at a suburban, mid-western community college. At the start of the class, there were twenty-one subjects whereas by the end there were seventeen. The purpose of the class was to have students acquire basic skills (*i.e.*, using search engines, learning HTML format) and general theoretical knowledge about the Internet. Each individual's purpose for taking the class varied. Some individual reasons for choosing the course included: fulfilling college requirements; learning how to create a Web page; enhancing Internet and computer skills for career; and learning the Net for fun.

Generally, the students in the class varied in age, ethnicity, race, socioeconomic status, and educational background; the class included four and two-year college students, retired students, and students who were already well into their careers. The majority of the class was white however, two Chinese women, a Middle Eastern male and an African-American male added diversity to the class. There were close to an equal number of men and women. As far as computer background, all had some limited experience with computers and the Internet; although, some had much more computer experience than others. For instance, one young man used a personal computer everyday at his job. He stated, "I use Microsoft word and programs such as Excel regularly." Other students, however, had no regular access to computers and little experience with the Internet.

Several methods of data collection were employed to achieve triangulation: Observation, semi-structured interviews, unstructured interviews, and documents (syllabus, class handouts). The methods were used to identify instructor techniques and behaviors that were either alleviating or exacerbating the students' Internet anxiety. While observing the class and conducting interviews, the focus was signs of anxiety such as cues of non-verbal frustration and fear cues of verbal frustration and fear, and cues of learned helplessness along with avoidance behaviour.

Miles and Huberman's three-part interactive model of data analysis was followed for this study. The first phase, called data reduction, includes choosing, focusing, consolidating, abstracting, transforming, and coding data gathered from observations and interviews. Part-II involves a data display which is concise, organized information such as in a graph or a chart. In essence, it is a summarizing image of the data. Lastly, part-III entails drawing conclusions and verification. It is at this phase that the triangulation of data is emphasized. Overall, Miles and Huberman's model is complete, detailed and well suited for the purposes of this study.

HTML (Hypertext Markup Language), SLIP (Serial Line Internet Protocol) connections and URLs (Uniform Resource Location) are just a few of the lexicons of the Internet/computer world that are part of learning in a beginners Internet class. A majority of students in this class had feelings of anxiety as they were introduced to a host of new vocabulary words and acronyms. A black male in his thirties explained that it was like, "learning a new language where you only understand pieces of information".

Arriving at the intended location on the Internet is not an easy task. There are often problems in limiting and narrowing searches, problems with documents that have been closed or never existed in the first place, and various technical blocks to finding the desired site. An older white male commented, "The Internet is like a maze-you can't look over the top—you have to keep backtracking all the time.

Sure it's very frustrating for some people". Several members in the class accepted the uncertainty of searches as a fact of life while others experienced moderate to extreme anxiety. During an interview, Diane, a thirtyish white female majoring in criminal justice, explained that, when she "gets stuck" and is not able to find the information she is looking for it, "eats me up". She said, "It sometimes gets to a point where I just don't want to go to class, I think I'm not going to make it through the day". In another interview, Newman, a young white electronic media major with a fair amount of computer experience stated that he felt extremely, "frustrated and betrayed" when he was not able to find the topic he was searching for on the Net. He added that it frequently caused him to avoid the Internet for a certain period of time.

Grand Success of Internet

At this time, technology has not caught up with the demand of the millions of people who want to go online. The Internet is clogged with more and more people, causing more and more busy signals and time delays. The "anger" and "frustration" several students felt, especially under time pressures in their own lives and homework due dates, was apparent in this class.

The Risks

A few students in the class expressed a diffuse fear of Internet failure. This was a generalized anxiety that they would not be able to negotiate the Internet, not be able to learn the content of the course if failed to complete the homework assignments and failed the tests. Jane, a young female education major with little computer experience and a dislike for technology, on her first day of class, repeatedly lamented, "Oh my God, I'm gonna fail this class". She dropped the class after a few weeks.

Low self-efficacy was the thread that ran through each of the four areas of Internet anxiety. A student who has low self-efficacy, as mentioned previously, judges his/her ability to perform a task as being low, feels less in control, and believes that low ability is internal. For each of the four areas of anxiety, there were examples of students in class blaming themselves for failing to perform. For instance, Diane exclaimed, "I think it's me. I think it's me. I think I'm not capable of getting where I want to get and when my friend Dana wasn't here today I felt lost... I got this panic feeling like, oh my God, he's gonna give us an assignment and I'm not gonna get to it in time". Even with Internet time delay anxiety which is basically a problem with the technology, Newman, felt it was his fault when he was stopped because of time delays. He explained that he felt, "stupid" and thought it must be his fault that he couldn't get to the information he wanted.

In another instance, Diane's friend Dana, who did not seem to experience much anxiety in the class, was not exempt from blaming herself for not being good at book learning (as opposed to hands-on learning with the computer) even though most of the class seemed puzzled after a lecture imbued with a barrage of cryptic text. From these examples and others, it was clear that, in this class, low self efficacy played a significant part in Internet anxiety.

The following section is a list of instructional techniques and behaviors, some of which mitigated the students' Internet anxiety in this particular class and some of which exacerbated the students' fears. All of the instructors' tactics related to one or more of the four areas of anxiety identified, in that he either addressed the concerns in these defined areas or failed to alleviate the students' anxieties as they related to the four categories of anxiety, including Internet terminology anxiety, Internet search anxiety, Internet time delay anxiety, and general fear of Internet failure. Additionally, many of the successful tactics used by the instructor addressed low self-efficacy concerns either directly or indirectly.

Progress of the Learners

The instructor emphasized the Internet as a "live Internet" which counteracted some of the false promises that people like Newman felt they had been bombarded with. The analogy of the Internet being like a living organism highlights the fact that it is not perfect, *i.e.*, sites come and go so people may not be able to find what they are looking for. The instructor also warned students about time delays several times which Newman said helped him to be less anxious because he knew it wasn't he who was doing something, "wrong or stupid". The instructor devoted class time to a brainstorming session about the pros and cons of the Net.

Systematic Learning

Technological Alterations: Programmed learning is one of the important invocations of the 20th century in the teaching-learning process. It is a self-instructional technique for providing individualized instruction or learning experience to the learner. In programmed learning, the subject matter or learning experience is logically sequenced into small segments. The learning experience is self-corrective.

The English writers prefer to use the term programmed learning and the American authors prefer the use of programmed instruction.

It is held by some educators that 'Gita' is the first example of programmed learning. They hold that the text of the 'Gita' has several ingredients of programming: initial behaviour, small steps, active participation of the learner, terminal behaviour, immediate feed-back and self-evaluation by the learner.

Several educators regard Socrates as the earliest programmer. Socrates used to guide his followers to gain knowledge by conducting them conversationally along a path from fact to fact and insight to insight.

Programmed learning emerged in the beginning of the 20th century from the efforts of American psychologists. E. L. Thorndike (1874-1949) was the first psychologist whose findings bear direct relevance to programming. Other important psychologists who have made significant contribution in the field are Sidney L. Pressy, Robert M. Gagne, Robert Mager and B. F. Skinner.

Programmed learning is related with the 'Law of Effect' as explained by Thorndike. Sidney L. Pressey, a psychologist of Ohio State University, is credited for developing in the middle 1920's practical machines which could teach as well as test. The teaching machines as developed by Pressey present a series of questions to a student and inform him immediately whether his response is right or wrong.

In 1943, Skinner and his two other colleagues started programming by teaching a pigeon to roll a small bowling ball by operant conditioning. By 1954, Skinner and James G. Holland devised the auto-instructional methods which have served the present generation as the basis for present work in programmed instruction. In Skinnerian programmed instruction, whether mechanized or otherwise, the learner is initially asked a question which he can easily answer correctly without any previous study of the particular lesson. The learner is taught by the sequence of questions. He is asked more and more as the lesson proceeds in very small steps. In 1955, Norman A. Crowder developed what he calls "automatic tutoring by intrinsic programming" as against "extrinsic programming" developed by Skinner.

Robert Mager (1958) gave a new concept known as "Learner Controlled Instruction" which is a kind of Socratic dialogue in reverse, in which the learner led the instructor. The instructor remained silent until the learner himself stimulated the instructor with questions that suggested the needed illustrations, demonstrations, practice or some other help.

Stoluron, at Illionis, aimed at developing a process which should provide for greater individualization by measuring needs and developing programmes that require a computer to assist instruction.

In 1962, T. F. Gilbert gave formalized expression of his technology of education called Mathetics. Pennington and Slack expressed in 1962 further detailed methods of preparing lessons from mathetic principles.

Programmed learning is a process of arranging material to be learned in a series of small steps designed to lead a learner through self-instruction from what he knows to the unknown of new and more complex knowledge and principles. A programme takes the place of a tutor and leads the learner through a set of frames of specified behaviour designed and sequenced to make it more probable that he will behave in a given derived way.

In programmed learning, it is said that the most efficient, pleasant and permanent learning takes place when the student proceeds through a course by a large number of small, easy-to-take steps. Wilbur L. Schramm (1962) lists the essential elements of programmed instruction as: (*a*) an ordered sequence of stimulus items, (*b*) to each of which a student responds in some specific way, (*c*) his responses being reinforced by immediate knowledge of results, (*d*) so that he moves by small steps, (*e*) therefore, making few errors and practicing mostly correct responses, (*f*) from what he knows by a process of successively closer approximation, toward what he is supposed to learn from the programme.

Following definitions provide a comprehensive view of programmed instruction.

Dale, Edgar (1962): Programmed learning is a systematic, step by step, self-instructional programme aimed to ensure the learning of stated behaviour.

Das, R. C (1993): Programmed instruction is a method of individualised instruction where each individual learns by himself at his own rate. Programmed learning consists of elements of new knowledge called 'steps' which are arranged in a sequence in such a way that a student can easily learn by himself.

Espich, James E. and William B (1965): Programmed instruction is a planned sequences of experiences, leading to proficiency in terms of stimulus response relationship.

Gulati and Gulati (1990): Programmed learning as popularly understood is a method of giving individual instruction in which the student is active and proceeds at his own pace and is provided with immediate knowledge of results. The teacher is not physically present. The programmer, while developing programmed material has to follow the laws of behaviour and validate his strategy in terms of student learning.

Jacobs and Others (1966): Self-instructional programmes are educational materials from which the students learn. These programmes can be used with many types of students and subject-matter either by themselves, hence the name "self-instruction" or its combination with instructional strategies.

Kampfer (1970): Programmed learning is a device which presents an exercise or a problem to a student, inducing him to respond; and revealing to him whether or not his response is correct.

Leith, G. O. M (1966): Programme is a sequence of small steps of instructional material (called frames), most of which require a response to be made by completing a blank space in a sentence. To ensure that expected responses are given, a system of cueing is applied, and each response is verified by the provision of immediate knowledge of results. Such a sequence is intended to be worked at the learner's own pace as individual self-instruction.

Luonsdaine Arthur, A (1964): An instructional programme is a vehicle which generates an essentially reproducible sequence of instructional events and accepts responsibility for efficiently accomplishing a specified change from a given range of initial competencies or behavioural tendencies to a specified or terminal range of competencies or behavioural tendencies.

Marke, Susan (1969): Programmed learning is a method of designing a reproducible sequence of instructional events to produce a measurable and consistent effect on the behaviours of each and every acceptable student.

Navi, N. S. (1984): Programmed instruction is a technique of converting the live instructional process into self-learning or auto-instructional readable material in the form of micro-sequence (the segments of subject-matter) which the learners are required to read, make some right or wrong response, correct wrong responses or confirm right responses and attain the complete mastery of the concept explained in the micro-sequences.

May, K. O. (1965): Educational programming is the scheduling and control of student behaviour in the learning process.

Smith and Moore (1962): Programmed instruction is the process of arranging the material to be learned into a series of sequential steps. Usually it moves the student from a familiar background into a complex and new set of concepts, principles and understandings.

Stolurow (1966): Programmed learning can be described as a process in which a teacher presents (*i.e.*, communicates), a subject matter to a learner so that he responds to it (*i.e.*, communicates to the teacher) the next item of information to be presented.

From the above mentioned definitions of programmed learning, following characteristics may be derived:

1. It is a method of individualized instruction.
2. In this technique, instructional material is logically sequenced and broken into suitable small steps or segments of the subject matter called 'frames'.
3. For sequencing a particular unit of the instructional material, the programmer has to pay due consideration to the initial or entering behaviour of the learner.

4. In actual operation, the beginning is made by presenting a 'frame'. The learner is required to read or listen and then respond actively.
5. Programmed instruction system has an adequate provision for feed-back.
6. The interaction between the learner and the learning material or programme is very important.
7. Programmed learning provides self-pacing to the learner.
8. Programmed learning provides for continuous evaluation.

Methods of Teaching

There are three basic types of programmed instructional material—The teaching machines, the programmed textbook and scrambled textbook.

The Teaching Machine: A teaching machine is intended to function as a private tutor. It is simply a mechanical devise or piece of apparatus designed to present to the student a sequential programme of learning activities comprising instructional items which requires the student to make an overt response and which provides the student with immediate knowledge of the accuracy of his response. It represents the practical application of laboratory technique of education.

Programmed Textbook: Each page of the programmed textbook consists of usually four or five panels. The student begins with the top panel on page one, responds to it, turns to page two to get his answers confirmed on the top panel, goes to the top panel on page three, responds to it, confirms the answer by turning the page, and so on.

The Scrambled Textbook: In a scrambled textbook, branching or intrinsic technique is used.

Principle of Small Steps: It is shown by experiments that even the dullest students can learn as effectively as the brightest students if the subject matter is presented to them in suitable small steps. When we divide the task to be learnt into very small steps, and ask the students to learn only one step at a time, then probably all the students will be able to learn one small step at a time and sequentially learn all the steps. It is a difficult task to climb a mountain but once steps are built even a child can climb the mountain very easily. This is known as the 'Principle of small steps'.

Principle of Active Responding: The second psychological principle is that the students learn better and faster when they are actively participating in the teaching-learning process. In our classroom teaching the teachers to ask a few questions and the students respond. But is not possible for the teachers to ask all the students to respond at each small step. A teaching machine text or a programmed text contains a large number of questions—one question at each small step and the students respond actively. The principle of active responding is used for the programmes. The teaching machines and programmes have proved to be superior because they provide opportunity to every learner to respond at every small step.

Principle of Reinforcement: Every response even approximately correct must be reinforced immediately. Delayed reinforcement fails to work. This is possible only when a teacher has to teach only one student at a time. The most ideal situation is when the teacher can cater to the needs of his students individually. But in classroom teaching this is hardly possible. No teacher, however efficient and sincere he may be, can reinforce each correct response of each of his students as soon as it is made in a classroom situation where he has to teach abut 40/50 students. The teaching machines and the programmes do the job more efficiently.

Principle of Self-Pacing: The programmed instruction is based on the basic assumption that learning takes place effectively if the learner is allowed to learn at his own pace. Therefore, a good programme of the material always takes care of the principle of self-pacing. A learner moves from one frame to another according to his own speed of learning.

Principle of Student-Evaluation or Student Testing: Continuous evaluation of the student and the learning process leads to better teaching-learning. In the programmed instruction, the learner has to leave the record of his responses because he is required to write a response for each frame on response sheet. This detailed record helps in revising the programme.

Important stages in the Development of the Programmed Instruction

Preparation: This is the first stage in the development of the programme. It includes:

(i) Selection of the topic.

(ii) Writing assumptions about learners.

(iii) Defining objective in behavioural terms.

(iv) Writing the entry behaviour (present status) of the learner.

(v) Developing specific outline of the content.

(vi) Preparing a criterion test.

Construction or Writing of the Programme: The programme is written under these heads: (*i*) Writing draft frames in a sequence, *i.e.*, from simple to complex. (*ii*) Editing the draft frames by a team of experts usually comprising a subject-matter expert, a skilled writer and the programmer.

Try Out Revision: It includes: (*i*) Trying out the programme on a few individual learners and finding out their reactions and making necessary changes in the light of reactions, (*ii*) Trying out the programme on a group of learners and making necessary changes on the basis of their reactions; and (*iii*) Trying out the programme in the field.

Evaluation: This implies finding the success or the failure after implementing a programme.

Civics-Class X

Introduction: This a programme meant for you for the study of salient features of the Constitution of India.

In this programme you will find paragraphs which are called frames. Study each frame carefully and write down what is required. Answers are given at the end. After stating you answers, check them. If your answer is wrong or you do not understand anything, you can again go back to the frame. It is not a test but instead it is a self-study programme.

Frame 1: The Constituent Assembly of India was set up under the provisions of the Cabinet Mission Plan to frame the Constitution of India which was formally adopted on 26th November 1949 and came into force on 26th January 1950. It took nearly three years to complete the work.

(i) What was the work assigned to Constituent Assembly?

(ii) Under whose provision was it formed?

(iii) When did our Constitution come into force?

(iv) When was it adopted?

(v). How much time did Constituent Assembly take to complete its work?

Frame 2: The Preamble of the Constitution has a great significance but is not a part of the Constitution. The Constitution was framed by the people of India through their representations. It stresses the fact that the reign of the land lies with the people of India.

(a) Is the Preamble a part of the Constitution?

(b) By whom was the Constitution framed?

(c) In whose hands does the reign of law of India lie?

Frame 3: The Preamble of our Constitution is as under:

We the people of India having solemnly resolved to constitute India into a Sovereign Socialist Secular Democratic Republic and to secure to all its citizens:

Justice, social, economic and political;

Liberty of thought, expression, belief, faith and worship

Equality of status and of opportunity; and to promote them all;

Fraternity assuring the dignity of the individual and the unity and integrity of the Nation.

In our Constituent Assembly this twenty-sixth day of November, 1949, do hereby adopt, enact and give to ourselves this Constitution.

Note—Three new terms—Socialist, Secular and Integrity were added to the original text of the Preamble when it was amended in 1976 with the 42nd Amendment.

The Preamble stresses the democratic basis of the Constitution by stating that the People of India gave to themselves this Constitution. It also states objectives like justice, liberty, equality and fraternity.

(i) Who has given the Constitution of India?

(ii) What kind of justice has been ensured by the Preamble?

(iii) What type of Republic is to be constituted?

(iv) What kind of equality has been given to its citizen?

(v) How many types of liberty can a citizen enjoy?

Frame 4: Another important feature of the Preamble is that the people themselves adopted and enacted the Constitution. Thus, the representatives of the people frame the laws of the country and they have the power to change or amend the Constitution.

(a) Who frames the laws of the country?

(b) What has the power to amend the Constitution?

Frame 5: The Constitution of India has many unique features which distinguish it from Constitutions of other countries. The framers of the Constitution freely borrowed ideas but took care to adapt these to the needs of the country.

The Constitution makes India a Sovereign, Socialist Secular, Democratic Republic. The word Sovereign means that India is completely free from external control. No outside power has the right to interfere either in her internal administration or direct her in the conduct of her foreign policy.

This was emphasised to ensure that India was no longer 'dependent' on the British Empire as she had been before Indian Independence Act 1947 or 'dominion' as she had been from 15th August, 1947 to 26th January 1950.

(a) What kind of status did India enjoy during 15th August 1947 to 26th January 1950?

(b) Was India sovereign between 15th August 1947 to 26th January 1950?

(c) What is the meaning of the word 'sovereign'?

(d) Does the Constitution of India have unique features?

(e) Did the framers of the Constitution borrow ideas?

Answers

1. (i) To frame the Constitution of India.

   (ii) Provision of the Cabinet Mission Plan.

   (iii) 26th January 1950.

(iv) 26th November 1949.

(v) Nearly 3 years.

2. (a) No.

(b) People of India through their representatives.

(c) People of India.

3. (i) People of India.

(ii) Social, economic and political.

(iii) Sovereign Socialist Secular Democratic.

(iv) Equality of status.

(v) Five.

4. (a) Representatives of the people.

(b) Representatives of the people.

5. (a) Dominion.

(b) No.

(c) India is completely free from external control.

(d) Yes.

(e) Yes.

How to Test?

Merits of Programmed Instruction:

1. A well-programmed instruction is a great thrust in the direction of individualised instruction, as it is tailored to the needs of the individual learner in the class.
2. It permits individual learner to progress at his own speed. An intelligent learner needs no longer to be bored or allowed to lose interest on account of his slow progress of other learners of the class. He can make progress as he is capable of.
3. Since a programme requires continuous response from the learner, it overcomes the inertia and passivity on the part of the learner.
4. The teacher can give explanation in the classroom if the error is common or he may arrange individual conferences on specific points.
5. Learning material in a programmed instruction is presented in such a way that learning becomes an interesting game and the

learner is motivated to meet the challenges set by his own capabilities.

6. Programmes are developed by experts. They are empirically tested and modified till they are standarised. A number of learners can use a single good programme and thus evade textbooks.
7. In programmed instruction the learner is immediately reinforced to correct his response and this reinforcement sustains the motivation of the learner.
8. The self-instructional technique presents material in which its complexity is simplified through the analysis of the subject-matter into small and more easily assimilated segments of information.
9. The introduction of programmed instruction is of great significance for developing countries which are set on the path of educating millions of learners and are short of teachers.
10. Good teachers are freed from the boredom of routine classroom teaching and they are in a position to devote more time to more creative activities.
11. The programmed instruction has been used more successfully in teaching the discernment of the logic of various disciplines and inspiring students to creative thinking and judgement.
12. Certain motor skills and intellectual abilities normally taught by frequent drills and rote memorisation can be very efficiently taught by self-instructional devices.
13. Self-instructional materials have been found to be very useful in the West in revolutionisng the social setting of the classroom. Problems of discipline have been solved and a new hope for eliminating emotional and social problems has been generated.
14. Programmed instruction enables the teacher to diagnose the problems of the individual learner.
15. The introduction of programmed instruction is very helpful in certain situations where human instructors are not easily available in the required number, for instance small schools in the isolated or hill areas.

Programmed materials have been severely criticised as a threat to replacing the teacher. It is also argued that there is too much emphasis in learning facts and very little emphasis on the mastery of principles and concepts.

Some critics of programmed instruction maintain that the user of a programme does not now where he is headed to.

They also point out that the learners are not aware of the organization and programmed instruction is unrelated to other aspects of instruction.

It is also argued that the programmed instruction material is very costly and only rich nations can afford it. It is also stated that the development and use of programmed instructional material require expert knowledge and training.

An average teacher finds it very difficult to make use of this device. As a result of experimental studies and research, following types of programmed instruction have emerged:

1. Linear or Extrinsic Programming
2. Branching or Intrinsic Programming
3. Mathetics Programming
4. Rules System of Programming
5. Computer Assisted Instruction (CAS)
6. Learner Controlled Instruction. (LCS)

The first three styles—linear, branching and mathetics are the basic formats. The rules system represents the deductive and inductive approach to teaching. The other two types, Computer Assisted Instruction (CAI) and Learner Controlled Instruction (LCI) are not the basic format of Programming. They are, infact, the ways and means of providing instruction. Here we have taken up only the basic type of programming.

Simple Method

B. F. Skinner is the originator of linear programming. It is also called a single tract programme. According to Skinner, a creature, a bird or a human being can be led to a desired behaviour by means of a carefully constructed programme consisting of small steps leading logically through the subject-matter from topic to topic, provided each step is reinforced by some kind of favourable experience or reward.

The increments in information which the learner is expected to absorb are small. The favourable experience or response increases the probability of the same response to occur again in the future. The process of rewarding the correct response to a stimulus increases the general tendency to give a response.

The sequence of frames and path of learning in programmed learning is systematic and linear. That is why, this type of programming is referred to as linear programming. Hence all the learners have to proceed through the same frames and in the same order. In a linear programme, learner's responses are controlled externally by the programmer sitting at a distant place. Hence linear programming is also termed as extrinsic programming. In branching programming, learner's response is controlled by the learner himself internally. It is, therefore, also called intrinsic programming.

Merits:

1. Immediate knowledge of results acts as a great motivator and releases anxiety and tension.
2. The smallness of the frames brings the sub-goals within the reach of the learner and thereby facilitates secondary reinforcement.
3. Repetition strengthens the responses and ensures learning.
4. Easy nature of the programme provides 'success experience' to the learner.

Limitations:

1. In linear programming, the learning process becomes quite dull on account of the following reasons (*a*) Subject matter is broken into very small pieces, (*b*) Responding is quite mechanical and restrictive, and (*c*) The learning process is quite slow.
2. The use of linear programming is limited to some subjects and topics.
3. Linear programming cramps the imagination of the learner and initiative for creative, integrative and judgement learning.
4. Linear programming encourages guessing.
5. Linear programming does not develop the discriminating power of the students.

Basic Method

Branching or intrinsic programming was developed by Norman A. Crowder (1954) an American technician. According to Crowder, branching or intrinsic programme is one which adopts to the needs of the learners without the medium of any extrinsic device such as a computer. It is not controlled extrinsically by the programmer.

Norman A. Crowder was a technician who was working in the United States Air Force. He was faced with the problem of efficiency of vocational training. His programme is based on intuition. His approach at the most is practical. This type of programme employs multiple choice response patterns. The learner is required to select one right answer out of several responses presented to him.

Merits:

1. Big size of a frame as well as the branching minimises unnecessary repetitions and responding, thus reducing the amount of learning time and fatigue.
2. The pitfalls and consequences of erroneous logic are usually explained in the remedial frames so that the learner not only gets the correct responses but also understands why some other response is not correct.
3. Instead of simple response it provides alternatives in the form of multiple choice.
4. Through its broad frames, branching programme provides for more freedom to respond and scope of choosing one's path of learning according to one's need. Thus, it helps in maintaining the interest and initiative of the teacher.
5. Branching programme is helpful in the development of the power of discrimination of the learner.
6. Branching programme helps in the development of creativity and problem-solving ability.
7. Branching is most useful in the areas beyond facts, definitions and basic skills.
8. The frames being of a large size contain a good deal of information and this may enable the programmer to enrich the style and expand his ideas.

Limitations:

1. The multiple choice questions provided in this programming may lead to guess work on the part of the learner and he may not understand the subject matter of the frame.
2. The setting of appropriate multiple choice questions suiting to the entire material of the frames proves a difficult task.
3. No branching method can provide infinite branching to take care of all possible needs of every individual student.
4. The cost of branching programme is very high when compared with traditional teaching approaches.
5. The branching programme is not suitable for small children as they are unable to express symbolisation.
6. The programme needs revision after every five years.
7. It is difficult to cover the entire subject matter of the curriculum in the stipulated time.
8. The diagnostic questions framed by the programmer may or may not suit the needs of the individual learner.
9. The programme cannot shape the behaviour of the learner.

Teaching Instruction and Programmed Instruction: According to Edger Dale, "Teaching' is a broad, vague, ill-defined term and instruction' is a purposeful, orderly, controlled sequencing of experience to reach a specified goal. 'Programmed instruction' is a sub-head under instruction and represents a more rigorous attempt to develop a mastery, over specified goals to secure 'insured' learning".

Programmed Instruction	*Traditional Method*
1. It is an individualised technique of instruction.	1. It is a group technique.
2. It is based on the teaching principles that have been known for years.	2. It becomes difficult to apply teaching principles in crowded classrooms.
3. It presents the instructional matter step by step in logical order.	3. It presents the instructional matter as a whole.

Contd.

Programmed Instruction	*Traditional Method*
4. The size of the unit of information presented to the pupils is a small bit of information.	4. The unit is a lengthy one. There is very little provision for response from the students in the form of answers to questions.
5. Immediate feedback is given to the learner.	5. The learner does not get immediate feedback.
6. Objectives are defined very clearly in operational terms.	6. Objectives are not well-defined and are usually vague.
7. The programmer prepares his programme with care and precision.	7. Very little preparation is made.
8. Programme is prepared in such a way that the student automatically participates actively by making responses continually.	8. The student usually remains a passive listener and the teacher himself does the summarising and reviewing.
9. A programme is developed empirically through a series of tryouts and refined gradually. Effective sequences students reaction of frames are retained and ineffective ones discarded.	9. It is usually found to be very difficult to modify traditional instruction.

ROLE OF TEACHER

Programme learning cannot replace the teacher. Any innovation in the school programmes and practices must remain in the hands of the teachers. The radio and TV did not displace the teacher. Similar is the case with programmed instruction. It is upto the enlightened teachers to take up the challenging task of preparing programmes. We have got a wide market. The programmes can be sold all over the country. A student who is convinced that he can learn better, achieve more with the help of this programme, will definitely prefer instead of buying this programme to buying a textbook. By taking up this challenging task we will not only help the cause of education, help our fellow teachers by setting them free from the routine task of information, giving help to the students to achieve more, but we will be helping ourselves also.

It may also be remembered that these gadgets can be used mainly in the cognitive field and possibly in the psycho-motor field to develop certain abilities and skills of the students as an individual. A teacher is something more than all these gadgets put together. He has to bring about socialization of the individual; he has to promote socially desirable attitudes and interests and mould the personality of the students. The effective domain is almost reserved for his care. At present the teacher is not able to devote his energy and time to this important task as most of his time and energy is consumed by his routine job as an information giver. We always talk of education for three 'H's'—the head, the hand and the heart. But it has almost remained a mere slogan. Programmed learning, teaching machines and other gadgets will set teacher free from routine work. These are labour-saving devices for the teacher so that he may function more effectively in a field of his own choice.

Technique involved in programmed instruction can be used in teaching different subjects. Teaching of mathematics, science, social studies and elements of Indian languages can be done effectively with the help of this new technique. The teacher has to formulate objectives of teaching a particular subject, undertake content analysis of the subject matter in the light of objectives, frame a chain of questions which will lead the pupils in the direction of the objective and present the questions to his pupils who are expected to try their hand at answering the questions independently. The teacher will have to play the role of a friend, guide and philosopher in the class when the pupils are engaged in solving the riddle and at the same time acquiring knowledge or skill. The question of class discipline may not arise as the pupils will be found busy doing the task assigned to them by the teacher.

The teacher will have to do remedial or corrective teaching as the weakness of his pupils will be located in the very act of learning. The pupil will also undergo a process of self-evaluation as he completes his work.

Role of the teacher in the changed context of Programmed Learning may be stated as under:

1. Teacher as an advisor in helping students in the selection of programme learning material.
2. Teacher as a discussion leader for focussing the attention of the learners on important points.
3. Teacher as a guide to clarify doubts and elaborate on various points asked by the learner.
4. Teacher as an evaluator of the learning outcomes.
5. Teacher as a consultant to the various agencies engaged in production of programmed material.

The programmed learning approach can be adopted in normal classroom teaching in the following ways:

1. A teacher can make use of the principles of programmed learning such as active responding, minimal errors and confirmation while teaching various subjects in the conventional manner.
2. A teacher can define behavioural objectives in advance of teaching.
3. A teacher can validate the instructional systems of a class in terms of the performance of learners immediately after teaching is over.
4. A teacher can regulate questions and answers. The answer of a learner can be immediately reinforced by informing or telling whether it is correct or incorrect.
5. A teacher can plan the entire instructional programme of a classroom and can treat the terminal behaviour, the pre-requisite skills and content analysis in advance.

Komoski (1960) an expert has observed "Two thousand years ago the world's first public administrator, a gentleman by the name Quintilian wrote what might be called a handbook for teachers". In it he has one bit of advice which will serve as an excellent starting point for a discussion of programmed learning and its potential uses. His advice is: "Do not neglect the individual student. He should be questioned and praised".

Systematic Instruction

Programmed instruction is still in its infancy in India. Programmed instruction as an optional or elective paper has been included at the B. Ed./M.Ed. level in a few universities in India. It also forms a part of the paper of Educational Technology/Educational Innovation. However, as regards its classroom use, it may be observed that it is almost nil. As far back in 1966, the Kothari Commission suggested to develop programmed material in different subjects to test the suitability of the technique in Indian conditions. An Association of Programmed Instruction has been formed to coordinate the research being done at different centres in the country. The association also disseminates the information on new studies through its journal issued from time to time. The National Council of Educational Research and Technology has also done some work in the field. In spite of all these efforts, it may be stated that the application of programmed instruction has not yet made an appreciable impact on our classroom teaching. Our methods of teaching still remain traditional, by and large.

6

COMPUTER FUNDAMENTALS

THE MEANING

A computer can be defined as an electronic device which can process large volumes of data using pre-defined instructions set by a human being with high speed and accuracy.

THE EXPLANATION

Human beings supply data to the computer which are required to be processed. Not only that one has to supply the instructions required to process the data. The computer remembers both the data and instructions, does the processing when asked to, and gives us the result in the specified form. Historically, computers used numbers as the primary form of input data. With the advent of technology, the definition of data has expanded. Users can now input textual (letters and special characters), audiovisual (music, video, voice, pictures), and physical (light, sound, temperature, pressure) as well as numerical data. Computers were originally invented to act as fast calculating machines. A calculator also helps in computations, so it can be considered similar to a computer. However, the scope of application of a calculator is limited. Normally, a calculator is used as a handy instrument to carry out small individual calculations and is operated by the user through special keys marked with symbols of mathematical operators. In contrast, a computer can store a series of instructions and a huge amount of data in its memory and can process a complete job for its users. Let us look into the building block of computers.

Importance of Von Neumann: Most of today's computer designs are based on concepts developed by John Von Neumann referred to as the Von Neumann architecture. He identified five units to perform operations on the data, namely INPUT, MEMORY, OUTPUT, ARITHMETIC AND LOGICAL UNIT (ALU) and the CONTROL UNIT. Traditionally Arithmetic Logical Unit and Control Unit are built together. These two together are called Central Processing Unit (CPU).

The diagram shows basic structure of a conventional Von Neumann machine.

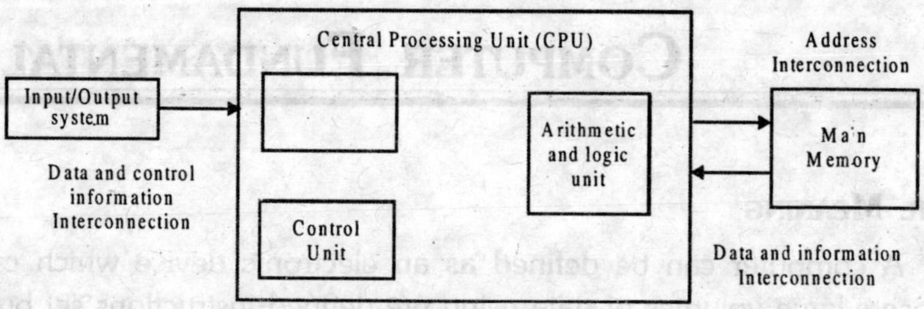

Figure : Structure of a Computer

Let's try to answer the following question:

How many units a Van Neumann Machine consist of?

The answer is quite simple. It consists of five units.

A problem with Von Neumann machine is that instructions are interpreted and data are processed and stored in a single sequential stream. This is referred to as *Von Neumann bottleneck.* Since there's only a single channel to carry instructions from primary storage to the control section, and since only data path is available between primary storage and the arithmetic-logic section, the processing speed is limited by the speed of the circuits in these channels.

Benefits and Brickbats

Advantages: A computer usually has

1. Very high speed
2. Large storage and retrieval capacity
3. Accuracy in calculation

4. Versatility in application
5. Diligence

Let us discuss the following characteristics of a computer in detail.

Speed: All the operations in a computer are caused by electrical pulses. We all know that electricity travels at the speed of light, computers too can perform innumerable operations in just 1 sec. The processing speed of a computer is generally measured in *nanoseconds.* By the time an average person takes to read the contents of this page, the earliest model of the IBM microcomputer would add together a million numbers and still have some time to spare! You can imagine what would be the speed of the latest models of computers running 100 times faster!

Storage: Computers have very large storage capacity. The contents of around two hundred sheets of A4 size paper can be stored on a small floppy disk, which is not even half the size of one sheet of paper. What would typically need large storage cabinets can be stored on a small floppy disk, which is not even half the size of a lunch box. Famous reference works like the "Encyclopedia Brittanica" and "Oxford Dictionary of English" occupies many large shelves in library. But today, they are stored in optical disks that are less in both size and weight. While this in itself is impressive, computers also provide very quick and easy access to all the data. Apart from storing text, today's computers are also capable of storing picture and sound in digital form, which means that even movies and music can be stored and sent to places quickly in the form of CD.

Accuracy: Computers have many check circuits built in and so normally they do not make mistakes. Errors in computing axe generally due to human negligence rather than technological faults. We have to always remember the fact that a computer is unintelligent and has to be given precise instructions and correct data to carry out its operations. If the instruction is faulty or might mean more than one thing or if the data is wrong, the result coming out of the computer too would be wrong. This phenomenon of wrong output due to wrong input of instructions and/or data is termed as Garbage in Garbage Out or GIGO in computer jargon.

Diligence: Computers, being machines, do not suffer from fatigue and lack of concentration. If five million calculations have to be performed, a computer can perform the 5 millionth calculation with same accuracy and speed as it performed the first calculation.

Versatility: Computers can perform a wide range of jobs with speed, accuracy and diligence. In an organization, it is quite likely that the same computers are used for diverse purposes such as accounting, generating pay-slips, keeping track of manager's appointments, play games during the recreation hours.

Limitations: Few limitations of computer are:

1. Cannot think
2. Cannot learn by experience
3. Cannot take independent decisions. Requires human instruction to take a decision.

Glance on the Past

The evolution of computer has passed through different stages before it reached the present state of development. The evolution probably started from the 3500 BC when human being first started to learn to calculate with a calculating machine named Abacus. Thus, the evolution of counting system from Abacus to modern Microcomputer is the result of continuous human effort in search of a more versatile and efficient machine.

Evolution of Abacus

Abacus seems to be the earliest calculating machine, which was developed by the Chinese. Abacus consists of beads, which can move, in wires. The wires represented the column. The right most column represented the unit, the next for ten and so on. Number were represented by moving the beads at appropriate column. Abacus could perform simple addition and subtraction.

Contribution of Pascal

Blaise Pascal, the French mathematician, laid the foundation of automatic computing. Blaise Pascal used his adding machine (1662) simply for addition and subtraction. The machine consisted of gears, wheels and dials. With this, calculation could be performed by dialing these series of wheels carrying the number 0 to 9 around their circumference. The wheel was arranged in such a manner that each wheel had 10 segments. This machine was later improved by Gottfried (Germany) to perform multiplication and division.

Charles Babbage and His Engine: Charles Babbage, a professor of mathematics at Cambridge University, invented a machine called Differential Engine—which could evaluate accurately algebraic expressions and mathematical tables—correct up to 20 decimal places. Later, he developed analytical machine, which could perform addition at the rate of 60 per minute.

For technical limitations of that age, Charles Babbage could not see his invention operational. However, and surprisingly enough, today's computer has building blocks, similar to what was proposed by him. For this reason, Charles Babbage is remembered as the father of computing. With him, his disciple and co-worker, Lady Lovelace is remembered with reverence as the first programmer of the world.

In 1934, Harvard professor H. Eiken developed an automatic calculating machine, which was called Mark-I digital computer. It's internal operation was automatically controlled.

The basic drawback of the mechanical and electromechanical computers were:

- The speed was limited due to the friction/inertia generated by the movement of components.
- The data movement was quite difficult and unreliable.

The designers were looking for better switching devices, storing of the mechanism and reducing moving mechanical parts. With the advent of Triode vacuum tubes, a better switching device of earlier years, designers shifted to designing electronic computer.

Importance of ENIAC: While Mark-I was the first electromechanical computer, the first electronic general purpose computer was built in 1946 by a team of professors at Moore School of University of Pennsylvania and was called ENIAC. The ENIAC ushered in the era of what is known as first generation computer. It was a bulky machine consisting of 18,000 vacuum tubes, 70,000 resistors, 10,000 capacitors, 60,000 switches and a large size occupying 1500 Sq. feet of space. It was a very fast machine as compared to its ancestors and could perform 5000 additions or 350 simple multiplication in just one second.

The main features of ENIAC can be summarized as:

- ENIAC was a general purpose computing machine where vacuum tube technique was used.
- ENIAC was based on decimal arithmetic.

Role of UNIVAC: In the year 1946, Eckert and Mauchly founded their own company and began to work on the Universal Automatic Computer, or UNIVAC. In 1949, Remington Rand acquired the Eckert-Mauchly Computer corporation and in early 1951 the first UNIVAC-1 became operational at the Census Bureau. It was followed by EDVAC (Electronic Discrete Variable Automatic Computer) which had a storage capacity of 1024 words of 44 bits each.

PROGRESSIVE STAGES

The word "Generation" for computers, indicates a step in technology. Every step includes a major change in the components used for constructing a computer. Originally the term 'generation' was used to distinguish between varying hardware technologies (in some cases software also).

First PHASE: The computers produced between 1942-1955 are called the first generation computers. They were extremely large with low reliability. They used *VACUUM TUBES* in their circuitry which were fairly large and generated considerable heat, needed lot of space to install and required special air conditioning system. Storage technique and the use of punched cards for input output were primitive. The language used was the machine language, whose knowledge was restricted to a few individuals. And above all, they were very expensive due to the increased installation and maintenance costs. *IBM-650* was, however, the most popular first generation computer and was introduced in 1950 with magnetic drum memory and punched cards for input and output. It was intended for both business and scientific application.

Disadvantages: The first generation computers had the following unfavourable features:

1. Too bulky in size and expensive
2. Unreliable for excessive heat generation
3. Air conditioning requirement was too much due to heat generation
4. Non-portable

Second Phase: The Second Generation computers used a more advanced technology by replacing the Vacuum Tubes with TRANSISTORS. A transistor is a two state device made from silicon. These components were smaller in size, easier to manufacture, less power consuming, cheaper and more durable. The processing capacity and the speed of operation of the components built from transistor were considerably increased. Computer storage technique improved with the use of Magnetic Disks. The machine language was replaced by high-level language like *FORTRAN*. One of the main computer series during the time, was the IBM 700 series. Each successful number of this series showed increased performance and capacity and reduced cost.

Advantages: In comparison to first generation computers, the second generation computer had the following favourable features:

1. Smaller in size as compared to first generation computers
2. More reliable
3. Less heat generated
4. Faster computational speed

Disadvantages: Some unfavourable features were also seen in this generation:

1. Air conditioning required
2. Frequent maintenance required
3. Commercial production was difficult and costly

Third Phase: The third generation computers employed integrated circuits in which all the elements of an electronic circuit were contained in a tiny silicon wafer. The Integrated Circuits *(IC)* based on the Small, Medium and Large-Scale Integration (LSI)-technology replaced the individual Transistors in the third generation computers. Each Large Scale Integrated Circuit had hundred or more components packed into an assembly. The LSI technology led to the development of very small but extremely powerful micro computers. Integrated Circuits IC refers to the miniaturization of electronic circuits—such that hundreds of components are formed on a small chip of silicon. This chip is able to perform a variety of functions that in the past required several different electronic components. The third generation computers are much cheaper and more reliable than the second-generation computers. They are faster with more capacity and allow connection of a wide variety of peripherals, particularly magnetic disk units.

Advantages

1. Smaller in size as compared to previous generation computers
2. Even more reliable than second generation
3. Lower heat generated than second generation
4. Reduce computational times
5. Low maintenance cost
6. Comparatively portable
7. Less power consumption
8. Commercial production easier and cheaper

Fourth Phase: Fourth generation machines appeared in the 1970's, utilizing still newer electronic technology, which enabled them to be even smaller and faster than those of 3rd generation. The fourth generation computers used VLSI (Very Large Scale Integration) technology. The VLSI devices have thousands of more components packed into an assembly. This technology led to the development of microprocessors, where an entire CPU circuitry is placed on a single chip. The circuits designed here, provided lower cost, reduced failures, smaller size, *etc.* Intel 4004 was the first microprocessor, developed by Ted Hoff. All modern day computers like IBM PCs, Apple Macintosh, SUN Sparc Stations, *etc.* fall under this generation of computers.

Advantages

1. Low cost machine
2. High speed
3. Large memory
4. Small size

5. Less power consumption
6. Floppy disk and CD is used as storage device
7. Hardware failure is negligible
8. Heat generated is negligible
9. Cheapest among all generations

Fifth Phase: Scientists are now at work with the fifth generation of computers. The world is moving towards the development of what one can call 'Super Large, Scale Integration' – which in turn will compliment and improve speed, miniaturization and cost reduction. Development of ARTIFICIAL INTELLIGENCE to make computer function and take decisions almost like human being, implementation of expert systems, robots, intelligent programmes, speech synthesizers as well as the use of video disks and tapes for external storage media, are included in the fifth generation of computer innovation.

Generation of Computers

Year	***Generation Number***	***Types of Technology***
1940	1	Valves (Vacuum Tubes)
1950	2	Transistor
1960	3	IC (Integrated Circuit) with LSI (Large Scale Integration
1970	4	Microprocessor with VLSI (Very Large Scale Integration)
Late 1980's	5	Hardware and Software technology

7

THE FORMATION

Considering the main functions performed by a computer we can broadly classify the hardware components into the following two types named as:

1. Central Processing Unit (CPU)
2. Peripheral Devices connected to and controlled by the CPU

Let us define and describe them in brief before going into the details.

Definition: A Central Processing Unit (CPU) controls the execution of programmes and performs the calculations. The CPU is also called as the brain of the computer. This is because execution of any task by a computer is controlled entirely by the CPU. The CPU activities include ensuring all instructions for a programme are executed and in the proper sequence, the calculations and logical operations are performed correctly and the instructions and data are available for successfully executing a task.

For executing its jobs the CPU mainly uses three main interactive components. These are:

- Primary storage section
- Arithmetic and Logic Unit
- Control Unit

ADDED THINGS

Definition: Peripherals are the devices that permit a computer system to communicate with outside world. The peripheral devices allow the computer to receive data and instructions from the outside world and give the information to the outside world:

1. The input Devices (accept inputting data)
2. The external storage Devices (storing data)
3. The output Devices (displaying or storing data)

For executing a job on the computer, the computer has to be provided with the instructions and data for the instruction to act upon. These instructions and data are to be fed into the computer. This is achieved using input devices such as keyboard or a mouse. Then, instructions and data entered into a computer are retained in the computer memory as long as the power supply to the computer is on. Once the power supply is switched off, the contents from its memory are erased. We, therefore, require certain other peripherals which can store data permanently on them, irrespective of whether the computer is switched on or off. These peripherals are referred to as external storage devices and such devices include a floppy diskette, or hard disk. After, a computer has executed a task, it generally produces some information. This information may be displayed on the display unit or given in the form of a print-out via a printer connected to the computer. This print-out is referred as a hard copy. The information can be stored in an external media for future use also.

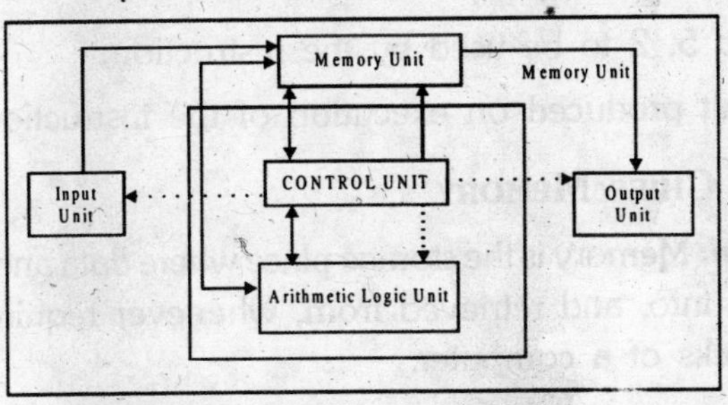

Block Diagram of a Computer

IMPORTANCE OF CPU

Central Processing Unit (CPU), consists of:

1. Primary Storage Section
2. Arithmetic and Logic Unit
3. Control Unit

PRIMARY STORAGE IN COMPUTER

Primary storage section, also called as main memory, may be compared to that portion of the brain where data and instructions get stored when a programme is being executed. The main memory is used to:

- Temporarily hold data received from input devices and keep them ready for processing.
- Hold data being processed and the intermediate results being generated therein.
- Hold the finished results of processing until released to the output devices.
- Hold the system software and application software in use (refer section "Software").

For executing the instruction X = 5 + 2 in Basic language on a computer running Windows 98 the following needs to be stored in main memory:

- The operating system (Win 98) which runs the computer.
- The interpreter for the Basic language to execute the instruction.
- The instruction to process the data.
- The data 5, 2 to be used by the instruction.
- The result produced on execution of the instruction.

COMPUTER'S CHIEF MEMORY

Definition: Memory is the storage place where data and instructions can be stored into, and retrieved from, whenever required by other functional blocks of a computer.

Description: Let us describe the main memory in terms of how it is organized. We can compare the computer memory with an almirah. An almirah is a storage place that has racks where items may be kept, so is a computer memory. A computer memory is provided with locations where the data can be stored. Each memory location is identified by a unique address. The locations store the data in the form of bits.

Normally each location can keep or store 8 bits (*i.e.*, a byte). The memory location addresses start from 0. Thus the other location is actually the first location of a memory. A group of 8 bits is called a byte. The total capacity of memory is expressed in terms of bytes. For example, a 256 byte memory would mean—a memory with 256 locations numbered from 0 to 255, and each location capable of storing a byte (8 bits). The memory inside a computer is normally very big, 655360 bytes or even more. To express such big numbers, we use terms like *Kilobytes, Megabytes, Gigabytes, etc.*

To be in tune with binary system a kilo in computer terminology means 21° or 1024 (in contrast to 1000 in MKS system). So a memory of 655360 may be expressed as 640* 1024 bytes or 640 kilobytes.

1 Megabyte = 1024 * 1024 bytes

1 Gigabyte = 1024 * 1024 * 1024 bytes

The specification of a computer is often expressed by size of the primary storage. The storage capacity is measured in Kilobytes (1 Kilo (1K) = 1024). A 128K computer is thus a computer having approximately, 1,28,000 locations for storage of data (each location holding 8 bits of data)

Example

In the following example, let us assume that a memory has 4 locations namely 00, 01,10, 11 (binary form; equivalent to 0,1,2, 3 in decimal; more on binary form in next section) as given in the diagram below:

		OPCODE			**OPERAND 1**			**OPERAND 2**	
Location	00	0	0	0	1	0	1	1	0
Location	01	0	0	0	0	0	0	1	1
Location	10	0	0	0	0	0	1	0	0
Location	11	0	0	0	1	0	1	0	1

Let us assume that the computer is so designed as to recognize its first instruction location. In this example, it is location 00. Let us also assume that bit positions 0 to 3 are taken as OP code by the computer, bit 4 and 5 as the 1st operand and bit 6 and 7 as the 2nd operand address.

The example instruction would eventually mean that the computer is supposed to add (OP code 0001) the contents of location 01 with that of 10. Location 01 and 1 does not contain any instruction but only contain the data, which will be used by the instruction stored at location 00. Examine location 01. It contains number 3 in binary form (11). Now examine location 10. It shows the number 4 in binary form 100.

Each computer has its own way of defining the instruction Syntax. Details of these can be obtained from the respective machine manuals.

Such an instruction in binary form with a series of 0's and 1's is called *machine code instruction*.

The operation explained is performed by the central processing unit by fetching the instruction, the operand addresses and the operands from the memory.

There are two operations involved with the memory. Storing operation is termed as write operation or writing in the memory; and the retrieval operation is termed as the read operation or reading from the memory.

Normally, a memory location contains 8 bits or 1 byte. The reading or writing of memory is also one location at a time normally. Though there are machines in which two or more bytes are read or written from a memory in a single read or write. The number of bytes fetched or written in a single memory reference is called the word size of the machine. There are also some machines in which, by special operation, a single bit in a location may be read or written. These are special cases.

Memory is broadly classified in two types:

1. Random Access Memory (RAM).
2. Read Only Memory (ROM).

Importance of RAM: The memory where both reading and writing can take place is called Random Access Memory (RAM).

Since access time in RAM is independent of the address of the storage location inside, the memory is as easy to reach as any other location and takes the same amount of time.

By technology, RAM may be of two types:

1. Static RAM
2. Dynamic RAM.

Place of Static RAM: Static RAM elements, once set to their values, do not loose them if the power supply does not go off. Once some value is stored, it will remain in the main memory till the power is on. These elements are costly and have a fast access.

Place of Dynamic RAM: In Dynamic RAM the memory elements have less retention period and lose its value unless it is refreshed. Refreshing is a process where the RAM is read and rewritten. This kind of RAM requires extra circuits for refreshing. These are relatively slow but less costly compared to Static RAMs. This kind of memory is therefore used in mass scale. Both of these are volatile, *i.e.*, the moment power goes off all contents of RAM are lost.

Use of ROM: There are portions in the computer memory where no writing is allowed. The contents are prewritten and can only be read. Such memories are called Read Only Memory (ROM). It is used for micro-programmes not available to normal programmers.

The term read only means that the storage cannot be altered by regular programme instructions. The information from the memory may be read out but fresh information cannot be written into it. The micro-programmes in read-only-memory may be used for a variety of purposes, but a common use is to hold a set of instructions that are needed frequently, for executing small, extremely basic operations, which are not otherwise available in computer circuitry.

Of the two parts of main memory, RAM is used as the work area where user's instructions and data can be stored and read. ROM contains those instructions and data, which will be required for the operation and is pre-written and would not change. For example, 'what the computer is supposed to do' when the power is put on and till there is an instruction from the user, may be written in the ROM.

The computer will be designed to carry out the initial instructions by reading them from ROM whenever the power is put on.

ROM can be classified technologically into 2 types:

1. Programmable ROM.
2. Erasable Programmable ROMs.

Programmable ROM or PROM are those in which the user can insert the contents of choice only once. Once the PROM is written, it can only be read and never be rewritten.

Erasable Programmable ROMs or EPROM is another kind. As the name suggests, it can be written many times by a special process. These are very useful for design and development.

Memory and the Speed

The fastness of main memory is expressed by memory access time. It is the time taken by the memory to read from or write in a single location of the memory. Speed is expressed in nano or micro seconds.

Some Factors

Other than the main memory, there are different memory elements used in different parts of the computer. They are:

- Cache
- Register
- Secondary storage, *etc.*

Importance of Cache

Cache is a very fast memory placed in between the main memory and the CPU. We know that each instruction is to be fetched by CPU from main memory. Main memory is comparatively slower than CPU, so it holds CPU whenever CPU refers to memory. Cache memory acts as a buffer and fetches chunk of contents from main memory to be used by CPU. Cache memory is very fast and very expensive, so amount of cache memory in any system is usually much smaller than the main memory.

Importance of Register

Registers are temporary storage places and holds data, addresses, *etc.*, for doing operations on them for data transfer. There are many registers spread through the computer. Some of the more important registers are like Programme Counter (PC) for storing the address of the current instruction being executed, Accumulator (AC) for storing one of the operands being used and also for storing the result of the operation. A group of registers is kept exclusively for hardware operation. Another group of registers is provided for the programmers for better performance and ease of programmes.

Function of Secondary Storage

The main memory is volatile, *i.e.*, contents of memory are lost the moment power is put off. For reusing data and programme, it is essential to store them. Presently there are certain input output devices used for this purpose. Some can store huge amount of data and can retrieve them at will. No programme can be executed directly from them. Programmes always need to be sent to the main memory for execution. The Floppy, Hard disks and Magnetic tapes are kind of memory. They are larger, cheaper and slower than the main memory but can keep data intact even after power goes off. They are called secondary storage. In contrast, the main memory is called primary memory.

In respect to speed and volume of memory available in a computer the following pyramid shows their relative positions.

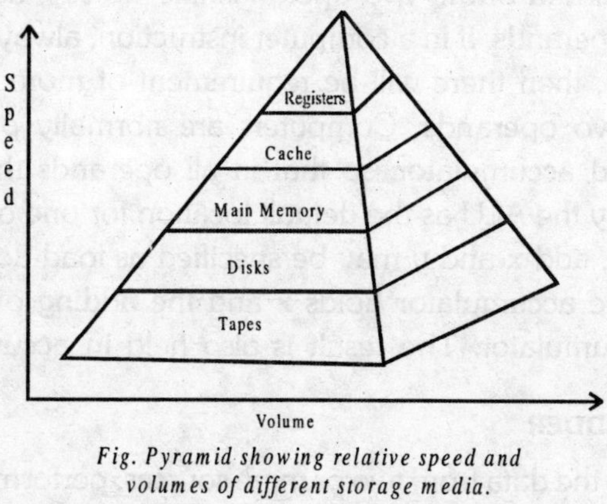

Fig. Pyramid showing relative speed and volumes of different storage media.

Importance of ALU: Arithmetic and logic unit does the operations on the data stored in the memory. This part of the CPU contains a large number of electronic circuits used to carry out arithmetical and logical computations. These computations are carried out with the help of special purpose storage locations, called the Registers.

There are basically three types of registers in an ALU:

1. Storage Registers
2. Accumulator
3. Adder

Different Registers

Temporarily holds data during execution of programmes. These are helpful in getting work done by a computer. For example, suppose data stored in two locations are needed to be interchanged. Then a third entity is required to hold the data temporarily. If the two locations are named x and y, then one can temporarily fetch x and keep the data in z. Then y may be fetched and transferred to x. In the third step, z may be copied to y. This third entity z may be a third location or a storage register. In a typical computer storage register operations are faster than memory operation. That is the reason why storage registers become handy.

Role of Accumulator

Holds the result of the calculations until needed later on. Operations are mostly carried out in two operands like we say, add x to y. Here x and y are operands. If in a computer instruction, always two operands are provided, then there will be requirement of more number of bits to express two operands. Computers are normally provided with a register called accumulator, so that in all operands this accumulator is assumed by the ALU as the default location for one of the operands. For example, add x and y may be specified as load accumulator with x, add y. The accumulator holds x and the adding of y and x takes place in accumulator. The result is also held in accumulator.

Role of Adder

Receives the data from two or more sources, performs the arithmetic and transfers the result to the accumulator.

The Adder is an important component in ALU. In any case, a computer performs all operations (subtraction, multiplication and division) through additions. Subtraction is done by a complimentary addition (a - b = a + (-b)), multiplication may be done by repeated addition and in some cases division may also be done by repeated subtraction.

ALU also has a counter to keep track of how many additions and subtractions have been carried out.

Importance of Control Unit

The Control unit contains that portion of the CPU, which controls each operation while a programme is being executed. You must have noticed that many times you keep a book open with your eyes fixed on it, but you do not read a single word. This is because the control unit of your brain was not allowing your eyes to give any input at that time. Possibly the control unit of your brain was processing something else.

Control unit synchronizes and sequences operations of the CPU for ensuring correct execution of programmes (sequence of instruction to the computer to perform various tasks). In other words, this unit acts as the central nervous system for various components of the computer, without as such performing the actual processing.

Control unit has a special register called the programme counter, which contains the address of the next instruction to be fetched from the primary memory. In order to sequence the instructions, the Control unit goes through the following sequence of steps:

- Retrieve the instruction from the processor memory addressed by the programme counter. Increment programme counter.
- Determine what action is being requested by the instruction. Retrieve the appropriate data from the processor memory.
- Direct the CPU components to perform the operation indicated by the instruction.
- Determine if the operation was carried out properly.
- If not, flag the error message to the user, through the output device.
- Else, store the result in the CPU memory, for further processing.
- Determine the location of the next instruction to be retrieved.

I/O Devices

Input Devices: The input devices are used to input data and instruction. Data and instructions must be entered into the memory of the computer to perform computation. This task is carried out by the input devices.

Devices themselves are unaware of what they input. Everything is input as raw data. It is interpreted as data or instruction by the Control unit.

Data read by input devices are of different forms, depending upon the form of input entity. Data from keyboard is entered in the similar way as typing, whereas a digitizer converts graphic data *(i.e.,* maps, diagrams, *etc.)* to make those understandable to computers.

Regardless of the type of devices, all input devices provide data to memory in binary form (codes of 0s and 1s). Working principles of each device is different. All devices will have an interface, which will enable the device to communicate with the system.

In brief, you can think of the generic functions of an input device as:

1. It accepts (or reads or inputs) the data from outside world.
2. It converts the data into the corresponding ASCII code or binary form.
3. It sends data in binary form to the computer for further processing.

Let us examine some standard input devices.

Some examples of input devices are:

1. Keyboard
2. Pointing Device:
   (a) Mouse
   (b) Joystick and Tracker Ball
3. Scanner

Function of Keyboard

A keyboard is the most common input device. It is used to input data manually by typing. The computer keyboard is like a typewriter keyboard with some extra special keys called function *keys and control keys*—which can be programmed by the user according to his needs. This means that if you write a programme, you might tell the computer to perform a specific action when a function/control key is pressed.

The common keyboard available is known as QWERTY keyboard. This name has been associated with this because of the sequential arrangement of these keys on the first character line of the keyboard.

The keyboard is connected with the main system through a cable. When a key is pressed or released, unique scan code is sent from the keyboard electronics to the system. The keyboard interface electronics generates the corresponding ASCII code for the key pressed.

Use of Mouse

Mouse is basically a pointing device. This is also an input device, but instead of sending characters, it sends the coordinates of the point on the screen on which the associated cursor is placed. A cursor is a specially illuminated character on the screen, which shows the place of selection for data entry or menu selection or a similar kind of thing.

The mouse is placed on a flat surface and is moved by the user on this surface. It has a roller underneath, which when moved, sends signals to the interface electronics to move the cursor the same way as the mouse is moved. There may be two or three buttons on the mouse. By pressing a button (normally the left button, though they are programmable), the user can point the selection to the system. The right mouse button is normally used for displaying and selecting pop-up menus (pop-up menus are special purpose menus and used only in specific context of a windows programme).

The mouse may be used to draw diagrams on the screen, to select one from a group of choices, or to place the cursor at any specific point on the screen. It cannot be used to enter data, the way it can be done using a keyboard. A mouse is a graphical user interface, namely WINDOWS.

A mouse is a must these days as it speeds up navigation in Graphical User Interface (GUI) packages and most packages today are GUI based.

Applications of Computer

Both of these are again pointing devices and are used for the same purpose as a mouse.

In joystick, a stick can be moved right, left, forward or backward. The electronics in the joystick measures the movement of the stick from its central position and sends the information for processing. The joystick is normally used for games programmes. The trackerball has a ball which can be rotated by hand in any direction. The cursor moves accordingly. It is normally used in medical computers like brain and body scanners, *etc.*

Importance of Scanner

The scanner captures the entire information of an image directly from the source (generally a page) and stores it in graphic format for displaying back on the screen. A scanner first illuminates the page into an optical image, which is then converted into a digital format, for storage by the computer.

The graphic image can then be processed using a suitable software. For, *e.g.*, if an image is stored as bitmap (BMP) file, it can be modified using MS-Paint.

Devices for Getting Results

The following functions may be attributed to the output device:

1. It accepts data in binary form from the computer, which are normally not understandable by human beings.
2. Converts the coded data to human acceptable form.
3. Outputs the converted result to the outside world.

Some of the commonly used output devices are:

1. Video Display Unit
2. Printer
3. Plotter

Device for Display

A TV like screen connected to the computer is called the Video Display Unit (VDU) or Monitor. It works on similar principle as a TV and displays graphics and characters on its screen. The screen consists of several tiny points (called picture points or PIXELS), which can be illuminated by highly focused electron beam. The beam scans the screen 24 to 30 times a second. It starts the scanning from left top corner, traverses to the right most point in the horizontal line slightly drooping down, then moves quickly back to left and traverses the next line. The process continues till it reaches the rightmost corner of the bottom most line. The scanning then starts all over again. If we draw the traverse path of the beam on the screen, the lines formed are called the Raster lines. Hence, these devices are also called as Raster Scan Devices. As the beam is always moving, the screen is coated with phosphor to retain the light. Otherwise the picture would fade away with the display time.

While traversing, the beam either illuminates a pixel or does not. These tiny dots are illuminated in such a pattern that we recognize the characters or the objects. A VDU may display colour too. In that case there are three electron beams to illuminate the Red, Green, Blue points on the screen.

VDUs may also be built by other technologies like Light Emitting Diodes, Liquid Crystal Diodes or Plasma.

PRINTING DEVICE

Printers are popular output devices. The output information is in a permanent readable form and is called the hard copy. Printers may be classified as:

- Character Printers
- Line Printers
- Page Printers

A character printer prints one character at a time, example, the *Dot Matrix printer.*

Line printers would print one line of text at a time. Example, chain or drum printers.

A page printer would print a complete page at a time. Example, Laser printer.

Printers are also classified as:

- Impact Printers
- Non-impact Printers

The impact printers use electromechanical mechanism to cause hammers or pin stud, to strike against a ribbon and paper arrangement. Dot-matrix, Chain, drum printers are Example of impact printers.

The non-impact printer does not have any mechanism to strike. It uses the chemical, thermal, electrostatic, laser beam or inkjet technology for printing. Non-impact types are normally faster than the impact printers.

Importance of Plotters

These are output devices and are used to produce high resolution graphics and drawings by the computers. The plotters use ink pen to draw. Colour pens are also utilized.

Various Devices

There are several devices used as secondary memory for computers and they can store and retrieve data in machine readable forms. These are useful for storing results temporarily, or for transporting data to other computers, or for storing data for a long period of time. The commonly used secondary devices are Floppy Diskette, Magnetic Disk and CD ROM drive.

Use of FDD

The floppy disk is made up of flexible plastic material. This plastic base is coated with an iron-oxide recording substance. Data is recorded as tiny invisible magnetic spots on this coating. Each floppy diskette is packaged in protective paper or plastic envelope. These are mainly used for the purpose of backup and transportability of data, for small data volumes. The floppy commonly used these days is 3% inch in diameter and has a storage capacity of 1.44 MB.

When formatted the floppy gets divided into Tracks, and tracks get further divided into Sector. Tracks are concentric circles and sectors are a small portion of a track, where the data gets physically recorded.

Sectors have their own capacity in terms of storing the data, which depends upon the type of floppy. While storing the data, if the sector is full, the data storing is done in the next sector and so on.

Some floppies used are double sided. Double sided floppy does not mean that you can reverse the diskette and read it, the way you do it for cassette tapes. Double sided here means that the electronics in the computer is placed in such a way, that at one go itself the data can be stored or retrieved from both sides of the magnetic material of the floppy. The process is transparent to the user. This type of floppy is double sided having 80 tracks per side and 18 sectors per track. Total capacity of floppy is 1.44 MB/2.88 MB.

Role of Magnetic Disk

A magnetic disk is an assembly of thin, circular metal plate coated on both sides with a magnetic material. Since, the plates are made up of metal, it is also known as hard disk. A disk pack consists of a number of these disks, three or more, mounted on a central shaft, which rotates at the speed of 2,400 or more revolutions per minute (rpm). Each disk or surface consists of a number of invisible concentric circles called tracks. A set of corresponding tracks in all the surfaces is called a cylinder. Outer surfaced of the top and bottom disks cannot be accessed. All the disks of a disk pack move simultaneously, in the same direction and at equal speed. Even though the length of each track is different, the number of characters stored in each track is same usually.

A hard disk having 6 disks with 100 cylinders, 50 sectors per cylinder and 500 bytes per sector can store,

500 bytes x 100 tracks x 10 surfaces x 50 sectors

= 25,000,000 bytes

@ 25 MB

The storage capacity of hard disks generally ranges upto several Giga Bytes (Ex: 4.3 GB, 9.14 GB 18.4 GB (Gigabytes).

Uses of CD-ROM: CD-ROM (Compact Disk Read only Memory) is an optical storage device which uses laser beam for reading and writing information. A CD-Rom is normally formed from a resin named polycarbonate which is coated with aluminium to form a highly reflecting surface. The information is stored on the CD-ROM by creating microscopic pits on this reflective surface with the help of a laser beam. A coat of clear lacquer is applied on the CD-ROM surface to protect it from dust and scratches.

The CD-ROM available these days have storage capacity 650 MB and more. Information on the CD-ROM is written using a CD writer. Information once written into the CD-ROM cannot be erased. However rewritable CDs are also available. In a rewritable CD, data can be written and deleted several times, as we can do with floppies. These CD-ROMs serve as excellent back up devices.

Functions of a Software

We have defined programme as a sequence of instructions to solve a problem using computer.

The sequence of operations may also be listed in steps for easy understanding. This step is called the Algorithm.

An Algorithm is defined as:

A set of well defined rules for solving a problem in a finite number of steps.

Let us understand with the following example:

Example

"Given two numbers x and y, find z = x + y".

To do this, one may write the steps as,

Step 1: Read the first number x

Step 2: Read the second number y

Step 3: Let z = x + y

Step 4: Print z

8

COMPUTER ASSISTED INSTRUCTION

Assuming that the societal changes wrought, by microelectronic technologies would have greater implications for changes in educational programmes than would the instructional potential of these technologies. Second, since future scenarios about life in the developing computer culture will, in the final analysis, be written by individuals, the capacities of schools to enhance self-realization will become increasingly significant. Third, all of the important issues related to uses of microcomputers ultimately reside beyond technology in human and social domains. Finally, the objectives of computer literacy programmes cannot be permanently fixed; they change as societal values and technology change.

DIFFERENT THEORIES

At a general level, differences in the conceptual and practical meanings of computer literacy exist within and across nations, at a specific level, they are found in the multiple programmes offered in given school. Diversity expresses itself through the scope and depth of computer literacy programmes, the objectives they seek, the outcomes they produce, and the places they occupy in curricula. Computer literacy programmes now found in schools are not easy to classify, in part because they are usually defined in general terms. In addition, those describing programmes often fail to distinguish clearly between concepts and their uses, and the resulting obscurities make classification difficult. Nevertheless, the categories of literacy elaborated below do help document and clarify the diverse expressions of current programmes.

Those who have acquired "operational" literacy can define the basic components of hardware and software and can demonstrate the machine-related skills needed to make computers work. Designed to introduce students to the computer, this approach is often implemented through daily exercises during a period of several weeks. When the exercises are effective, students understand such concepts as "monitor", "keyboard," and "central processing unit" and how the components labeled by such concepts work together with other parts to make computers function.

Students are also able to turn a computer on and off, load software, show proficiency at the keyboard, demonstrate the use of storage devices, and acquire "minimal" troubleshooting techniques. Operational literacy, then, emphasizes "hands-on" experiences and simple concepts. Directed at elementary understandings and skills, the approach demands neither extensive teacher training nor substantial amounts of instructional time. Nor does it require much change in existing curricula or major investments beyond those made in hardware and software.

The minimal demands of the approach may account for the fact that many school districts have opted for it in their initial literacy efforts. The approach, while limited, provides an entry into the world of the computer and a potential bridge to more significant learning experiences; more easily than other literacy alternatives; operational literacy can be made available equitably to all students. Learning produced through operational literacy, critics argue, are both trivial and subject to quick obsolescence. They might contend, for example, that teaching students how to turn computers on and off is just as trivial as teaching them how to turn television sets on and off. Both activities squander valuable time. Critics also argue that operational literacy outcomes are immediately out of date, hardware and software are changing so fast, they say, that today's operations are obsolescent tomorrow.

In addition, future computers, with their enhanced "friendliness," intelligence, and simplicity, will require only the most elementary skills. Why waste students' and teachers' time, critics ask, by providing trivial and obsolescent learnings? Is the assumption about obsolescence as sound one? Since students of technology typically underestimate, often substantially, the time required to realize new developments, a note of skepticism is in order. Put differently, projections made about impending achievements in intelligent and friendly computers by critics of operational literacy, especially the highly optimistic ones, are likely to be in error. If the new technologies have potential for enhancing learning, then a pertinent question arises—why wait years for more powerful computers while immediately closing the door on opportunities for all students to acquire potentially useful entry skills and concepts?

Is the "triviality" argument sound? If the learnings acquired through operational literacy offer no foundation for new learnings, they can be classed as trivial. However, if the elementary and first-stage learnings facilitate the attainment of more advanced ones, they are not necessarily trivial. To be sure, research is needed to determine if the "operational" learnings acquired in fact enable students to realize more sophisticated learnings. Further, do students from diverse backgrounds actually use the initial learnings to acquire more advanced ones?

Instrumental literacy approach to literacy computers are used as instruments for acquiring stipulated learnings or bodies of subject-matter. In Sweden, for example, leaders have used computers in high schools to teach mathematics and civics. They chose these subjects in parts because they "are taken by all compulsory school pupils, and they are included in most lines of upper secondary school". The approach enables all enrolled students to learn to use computers as they acquire content or skills; in addition, the computer can be used as an instrument for "modernizing" subject-matter teaching.

In the Swedish plan, for example, computer-supported projects produced, according to one observer, a variety of subject-related outcomes among which were new insights by students into mathematical relationships, into "connections" among subject matters, and into the analytical aspects of mathematics. In addition, teachers coordinated instruction in physics and mathematics more effectively. The Swedish approach reflects the instrumental use of computers in selected subjects. Some high school leaders have aspired to use computers in all subjects taught. Those taking clerical courses, for example, might acquire word processing skills; some social studies students, insights into environmental problems through simulations; science students, understandings of physics from computer-based demonstrations; language arts students, enhanced vocabulary and spelling skills, and so forth. In each subject, in other words, the computer would provide a tool for advancing learning.

DIFFERENT NOTIONS

The instrumental concept of computer literacy is not without its critics. Some detractors maintain that this alternative provides "low level" types of learning only. Others are skeptical about whether the quality of future software will exceed significantly that currently available in today's drill and practice programmes. On the one hand, school systems do not have the material and human resources to develop effective software, they maintain; on the other, the private sector, with its press to produce programmes for mass markets, will continue its "least common denominator" stance toward software. In describing the low state of software development, critics pinpoint a major difficulty.

However, the conditions they describe are not sufficiently compelling to justify the elimination of the instrumental approach to computer literacy. Information presented in this volume suggests that computer-assisted instruction, when compared to other approaches to the improvement of student learning, already offers one viable option to educators. Undoubtedly, future improvements in software will come slowly and incrementally, and the cost of hardware will diminish less rapidly than in the past; nevertheless, comparisons of CAI with competing strategies of instruction are likely to become more rather than less favourable over time. In the meantime, CAI provides students some computer skills and provides staff a stimulus and a medium for assessing traditional modes of instruction.

Algorithmic reasoning, perhaps the most controversial concept of computer literacy, is acquired through computer programming. A.P. Ershov, a member of the USSR Academy of Science, has called programming the "second literacy". Literacy in reading and literacy in programming both have their origins in technologies—the printing press and the computer. Both support basic functions; "communications work, the contemplation of nature and the struggle against it", and they "supplement each other, forming a new harmony of human mind". Ershov suggests that the "acceptance of books blew up the integrity of word and action; it separated the process of knowledge accumulation from its application".

Programming, with its emphasis on problem solving and on the clear comprehension of how to carry out what is to be done, helps rejoin, at least in certain types of tasks, what the book rent as under. To become programmers, students must make effective use of algorithmic reasoning, Kemeny writes:

We have to learn to think in terms of algorithms, which are precise recipes for solving problems or for accomplishing a task. A computer programme describes an algorithm; it tells us "how to get there". Although algorithms were used and studied in mathematics in the past, there is today a strong prejudice in favour of formulae-that is, achieved results. But computers are forcing us all, even mathematicians, to readjust our thinking.

Algorithmic reasoning skills, Kemeny admits, are not easily articulated. Nevertheless, he suggests that programmers need: (a) to be able to express thought in very simple language (the small number of concepts available in a language like BASIC makes the task a demanding one); (b) to explain to the computer how to solve the problem (c) to divide complex tasks into small, clearly understood components; (d) to break down a procedure into simple steps and seek the logical relations among them; and (e) to find programming malfunctions and "debug" them. While recognizing that programming can teach "procedural" thinking, Seymour Papert has stressed its potential for providing fundamental insights into mathematical and related abstractions.

By using algorithmic reasoning to make the computer do their bidding, students can gain special insights into thinking and into the nature of knowledge. Put another way, learnings gained from programming can help students become their own "epistemologists". Not only can programming develop generic reasoning skills, it can, its advocates contend, improve the quality of learning. Students, as programmers, are active and involved-preconditions for significant learning. Papert has also suggested that computers can bring constructive change to "sluggish and conservative" school systems. When teachers and students become skillful programmers, they move the posits, beyond passive contexts to richer and more self-actualizing ones. Some advocates of programming cast a critical eye on instrumental literacy because it places students in a passive role. As programmers, students can use computers in the active pursuit of their own learning objectives. If their responses are limited to those in programmes developed by others, their imagination and learning are sharply constrained.

Papert and Kemeny, as advocates of programming, speak and write as trained mathematicians. Algorithmic reasoning, an offspring of mathematics, can resolve precisely defined problems. It has repeatedly demonstrated its capacities to resolve technical problems associated with the development and use of microelectronic technologies. In fact, the expansive array of digital artifacts that lies at the heart of the computer culture is in one sense the product of algorithmic thinking. However, the task teaching students computer programming differs markedly from that of teaching them to address effectively the human, social, occupational, and civic problems created by microelectronic technologies. These latter problems lie outside the rational world of computer programming and in the emotion-laden contexts of political organizations, work settings, family milieus, and other locales.

Surrounded by conflicts, these formidably complex problems cannot be fully resolved through algorithmic reasoning. The contentions that all students should acquire computer programming skills, then, should be viewed critically. To make computer programming a literacy standard for all students reflects a questionable strategy. Even if the standard is one which can feasibly be met by all students-a proposition that is far from self-evident-major investments in time and resources would be required to produce a narrowly based set of skills.

Important Concepts

Nobie has identified "consumer literacy" "worker literacy," and "citizen literacy" as the major role-related concepts. Although writers do not always define these terms in the same way, the concepts do accurately reflect prevailing views about which roles are being altered by computers and which ones can serve as foci for developing computer literacy. Consumer literacy carries two meanings, one linked more to hardware and the other more to software. The first stresses that individuals should be effective acquirers and users of computers and associated hardware.

Since almost all families will eventually spend large sums of money on home computers and peripheral equipment, advocates of consumer literacy reason that schools should teach all individuals how to handle such expenditure wisely. The second meaning of consumer literacy centres upon the "applications" individuals will be making "as part of their strategies for information retrieval, communication and problem solving". Advocates of this view optimistically project rapidly increasing flows of information out of, into, and through home computers. Individuals using the expanding sources of information will need requisite knowledge and skills. They will need to learn, for example, how to enhance their education through electronic media, exchange ideas through computer networks, and use shared databases to solve individual and community problems. Schools, then, should provide students with the literacy skills required to function in tomorrow's microelectronic environments.

Worker literacy is defined, on the one hand, as the skills needed to perform specific jobs and, on the other, as generic cognitive and information skills. The first definition presumes that the trend away from industrial jobs and toward information jobs will continue and, further, that career and vocational programmes should teach individuals to perform such specific information jobs as computer repairing, systems design, and data processing.

The second definition of literacy presumes that future job requirements will always be changing and that schools, more than ever before, should provide students basic skills which they can apply as their jobs undergo change or as they move into new ones. Workers in information-related jobs, in other words, need higher-level mental skills than industrial workers traditionally have had. During the last decade, the impact of computer technologies on political environments is so far-reaching, some argue, that new approaches to citizen literacy are sorely needed. Today space and time no longer constitute barriers to national and international communication.

However, television and satellites are increasingly communicating political events through images, leaving it to citizens to search for the deeper political meanings behind appearances. Further, as the new technologies take over an ever-expanding number of government and business functions, they are posing far-reaching questions related to the privacy of individuals, equity of citizen access to the new media, and the role of computers in politics. All of these conditions, proponents of computer literacy argue, highlight the need for citizen literacy. Actual descriptions of well-developed programmes directed at consumer, worker, or citizen literacy are not as prevalent as are descriptions of other types of computer literacy programmes.

For one thing, they are more difficult to design and implement than are operational, instrumental, or algorithmic reasoning programmes. Since the definitions of role-related needs are still ambiguous, organized content to support instruction is not yet widely or conveniently available. Finally, support for the idea that schools should provide students generic or higher-order learnings for use in consumer, worker, citizen, and other roles is growing. Role-related literacy has not escaped criticism.

Advanced Technologies

Computer enthusiasts, some contend, have exaggerated the impact of the new technologies on job requirements, consumer demands, and citizen responsibilities. Levin and Rumberger, for example, suggest that the number of "information" workers required will be relatively small as compared to the number of janitors, cashiers, secretaries, and other service, workers needed. Others contend that school and community leaders have launched literacy programmes without adequate reflection. Some have blamed an "uncritical press" not only for exaggerating the need for literacy programmes but also for providing "free advertising in the guise of objective reporting".

Many educators have tended to accept the claims of computer enthusiasts uncritically. Thus, criticism is needed to develop more realistic views about the educational significance of the new technologies. However, the claims of the critics can be accepted as uncritically as those of enthusiasts. Stated another way, criticisms that lack logical or evidential support need to be viewed skeptically.

Critics of role-related literacy may be correct within an immediate time frame and incorrect within a longer-range time frame. Expanding computerization is changing consumer, worker, and citizen functions, even if slowly. Thus, today's critical messages about role related literacy not only may prove to be errant; even worse, they may reinforce outmoded patterns and stifle the creation of needed new ones.

Advocates of computer literacy, then, have stimulated considerable thought about the concept during the last decade. Witness, for example, the analyzes that increasingly dot the educational literature. By aggressively defining and diffusing multiple definitions of literacy and by skilfully linking them to diverse learning objectives, they have contributed significantly to the development and diffusion of literacy programmes. However, the mixed messages elicited by computers have slowed the diffusion of the concepts.

Some, seeing nothing new in computer technologies, indifferently dismiss them; in so doing, they rule out the "Whither" question. Others, sensing a computer-created "revolution," urgently call for major adaptations in school programmes and a thorough answer to the "whither" question. Given such disparate views, the first question facing those studying computer literacy is whether or not computer uses in schools indeed reflect another passing fad.

The indifferent, who "have seen it all before," believe firmly that the microcomputer is just another in a series of technologies without impact. They point to the high but disappointed hopes past educational leaders have had for programmed texts, televisions, and related innovations. They conclude that after a period of uncritical use the microcomputer will have an uninfluential ending as have past instructional technologies. Since it is only one more in a long series, they ask why one should assume that it has more than passing significance. The "passing fad" argument is erroneous because it is based upon a false assumption, namely, that the capacities and effects of microcomputers are similar to those of previous technologies.

Microcomputers in fact differ substantially from past technologies. Because of their greater versatility and flexibility, they can be put to many more instructional uses. In additions, they can help manage instructional processes; versatility access subject-matter from libraries, information utilities, and other electronic sources; and facilitate the development of programmes by students to meet their own objectives-feats that past technologies could not perform.

Computers differ from past technologies in a more fundamental way, namely, in their far-reaching effects on society. The impacts of programmed texts, for example, were confined to educational contexts. Even television, with its widespread influence, has not moved as deeply into society's institutions as has the computer. Microcomputers populate homes, work places, business enterprises, government agencies, social agencies, and other non-school organizations. Having spawned a powerful array of novel industries, computers are now stimulating new dreams and aspirations, redefining human machine relationships, shaping cognitive psychology, and altering modes of thinking.

As computers change society, they create new educational needs, as already noted. Microelectronic technologies, then, represent much more than new educational means, they are surfacing profound questions about the ends of education. Therefore, they cannot be blithely dismissed. The conclusion that the computer is not another passing fad does not imply that school purposes and programmes will change rapidly. Powerful barriers constrain technological change. The new technologies do demand much more learning and change on the part of school personnel than have past technologies.

The high costs of implementing literacy programmes will slow diffusions processes. The clashing tides of positive and negative opinion, which surround decisions about computer uses, will complicate adoption decisions. Finally, it is significant that much larger private-sector investments are currently being directed at 'offices of the future' than at 'schools of the future', and 'at functions performed in business, military, telecommunication and government agencies in contrast to educational ones".

The new technologies have evoked both dark and bright scenarios. Optimists, emphasizing effective uses, see computers producing beneficial outcomes; pessimists, emphasizing misuses, see destructive and even dire consequences. While enthusiasts make expansive claims, skeptics decry these claims and cast them aside. Hope, optimism, and enthusiasm clash with fear, pessimism, and skepticism. The result is bifurcated views. That these views impinge directly upon the 'whither' question can be shown with some specific examples. The first set of examples is linked to current computer uses; the second, to presumed future impacts of computers.

Sherry Turkle has referred to the "Computer as Rorschach". The phrase is an apt one; computers do tap deep human emotions and elicit contrasting perceptions about their meaning and significance. Applications of computers in the political arena, for example, bring dissonant views. Some examples relevant to citizen literacy can help clarify the point. Some have probed the impacts of the new technologies on individual liberty and privacy and have discovered an "electronic nightmare". Others, after studying national experiments in "teledemocracy," have concluded that the new technologies can help citizens influence public policy decisions.

In a similar view, some have observed that the power of "technological elites" constitutes a threat to democratic decision making; others see in the diffusion and use of microcomputers a new means for advancing decentralized communication and democratic action. Or while some argue that linear thinking can be effectively applied to political problems, others contend that intuitive thinking will be required to solve these problems. Some believe that expanding international databases will enable leaders to address worldwide political problems more effectively; others contend that leaders from developing nations will suffer even greater inequalities in political opportunities because they will be unable to access the data.

While the less sanguine stress that large technological systems, with "lives" of their own, cannot be controlled through political processes, the more cheerful contend that humans, as creators and users of these systems, can control their effects. Contrasting positive and negative views, such as those just noted, have implications for research. Inquiry is needed to shed light on actual in contrast to presumed outcomes of computer uses.

Some of the differences in view can be explained by the fact that scholars, in drawing their conclusions, use data about different but related phenomena as, for example, the computer's impact on a process (learning) in contrast to its impact on a value (privacy). However, a more fundamental problem is that the factual premises to support generalizations tend to be weak; effective research could uncover unreliable findings and false conclusions.

Perceptions about future consequences differ just as much as do those about immediate results. Thinkers often hold antithetical views, for example, about the long-rangé impacts of computers on the economic sector views which have implications for "worker literacy" programmes. Some, looking at the steady infiltration of robots into industrial settings, conclude that the result will be a permanently unemployed "underclass"; others, examining the impacts of past technologies on unemployment, assert that the new technologies will create at least as many jobs as they eliminate.

Some contend that the new machines will inevitably dehumanize work; others argue that computers, by taking over the routine, will bring shorter work weeks and that this, in turn, will lead to more educational, social and other humanistic activities. While the optimists believe that ongoing research will make it possible for robots to perform increasingly "intelligent" functions, the least hopeful maintain that research break-throughs in artificial intelligence will be extremely limited. The act that leaders and scholars hold such sharply divided views about the immediate and long-range significance of new technologies signals in still another way the powerful forces unleashed by computers. By surfacing conflictive and vital issues, the forces are stimulating and encouraging proactive throughout.

The precious discussion has stressed that computer literacy is a young concept, that educational leaders have defined and implemented it in diverse ways, and that sharply clashing views surround its uses. At this point an important issue arises, namely, what framework can be used to analyze the newborn, diversely, interpreted, sharply contested and complicated concept of computer literacy? Since the history of the concept is short and relevant research is scarce, analogies offer one framework. Thus, acquiring computer literacy can be compared with the learning of a foreign language or the attainment of "book" literacy. Book literacy offers one fruitful analogy for probing the "whither" question.

Notably, book literacy and computer literacy are both similar and interdependent. The various symbols on the keyboard used to write this chapter are immediate indicators of the reciprocal relationships between written literacy and computer literacy. Other symbols, which appear from time to time on the screen are additional reminders of the interdependencies between books and computers. These linkages underline a significant fact, namely, that individuals cannot acquire computer literacy without literacy in written language. Since one-fifth of the population of the United States is illiterate in written language, when more demanding measures are used, millions are locked out of computer literacy.

On-paper and On-screen Words

The linkages between computers and books, then, suggest an important policy generalization, namely, that when school leaders make commitments to computer literacy they should not do so at the expense of traditional literacy programmes. Those comprising today's large illiterate population tend to be members of disadvantaged social, economic, racial and other groups. Since as a rule they have had limited access to written materials in their homes, they have long experienced unequal learning opportunities. Such inequalities, which still persist 500 years after the invention of the printing press, offer another lesson to those concerned with policy related to computer literacy.

The opportunities available to students pursuing computer literacy will undoubtedly be more inequitable, at least for the foreseeable future, than will those available to students pursuing book literacy. Since computers and associated peripherals are much more expensive than books, educational agencies and parents will find it more difficult to provide children and youth the equipment needed to ensure equal opportunities. In addition, there are already public institutions that can diminish inequities in reading opportunities; however, public institutions to nurture computer literacy are not yet widespread.

If support for the concept of computer literacy continues, equity issues will increasingly find their way onto the agenda of school boards. Some educational leaders are already addressing the issues. Affirmative programmes directed at computer literacy during the summer, after-school scheduling to expand computer learning opportunities, and the use of such concepts as "student-access ratios" all reflect responses more random than widespread and more inchoate than well developed. The concept of "public information utilities" highlights the access question in the larger context of society. According enable all interested individuals to use large electronic databases to pursue educational, citizenship, and related purposes.

The implementation of this idea is at best a long-range possibility. It could not be realized until more advanced technologies are achieved, new information norms and habits are developed, and the transmission of computer-based information becomes much less expensive, nevertheless, the adjective "public" that stands before "information utilities" denotes an already emergent sensitivity is society to the equity issue. Given the very slow diffusion of reading skills during the last 500 years, public funds to support electronic information exchange will likely come slowly, if at all.

The first to achieve new skills are usually not eager to transfer them to others. Thus, when individuals, centuries ago, acquired written language skills, they became members of an elite group; as elites, they opposed the broad sharing of literacy skills. For example, the reading elite in England in the nineteenth century "viciously criticized" the idea of free libraries for the common people. Arguing that such libraries would cater to the "popular passion for light reading, above all for light fiction," the elite contended that the mass reading of such books would deter human progress, the common people would simply read the wrong things. It is possible that even today those who have acquired highly sophisticated computer skills tend to look down upon the use of games and drill and practice routines in schools.

Those opposing and those supporting the mass diffusion of literacy skills have confronted one another for centuries. Leaders on the side of the people, including the most disadvantaged, have prevailed, even though their struggle is not yet ended. Aided by powerful institutions, especially governmental agencies, they have repeated and acted upon a simple message, namely, that if individuals are to be enlightened and effective, they must be literate. If computers fulfil their educational promises, the struggle to eradicate inequitable learning opportunities will continue. As computer-utilities, and international promises, the struggle to eradicate inequitable learning opportunities will continue.

As computer-driven interactive cable, electronic universities, information utilities, and international networks are increasingly used, equity issues will become more salient. Since microelectronic technologies are more complex, expensive, and mutable than books, the equity issues they raise will likely be even more difficult to resolve than those associated with written language.

The five-century story about book literacy is broader than that of ever-expanding learning opportunities; it also encompasses a long struggle to define and redefine the goals of literacy in ways that reflect changing societal values. Thus, Patricia Graham, writing about literacy skills in the early 1980s, observed: "literacy... is what everyone needs in late twentieth-century America, both for their own good and for society's good". However, Horrace Mann argued, from a nineteenth-century perspective, that if students acquired literacy, vice would be reduced and morality would be improved a view that in today's climate would strike many citizens as incredible. In Mann's time, however, morality was such a salient value that school leaders could not easily ignore it. Given the secularized public schools of today, Mann's definition of literacy is no longer highly relevant. Definitions of computer literacy will probably follow in the footsteps of traditional literacy; they will, in other words, be temporary and impermanent.

Rather than, policy makers will need to seek definitions and objectives that mesh with the needs, of their time. Since the information society is much more mutable than previous societies, computer literacy objective will likely undergo more frequent changes. Given the many fearful messages about robots and related machines and their deep cultural roots, computer literacy decisions will also likely be more controversial. Witness, for example, the currency of the dark omens expressed a century and a half ago in Shelley's Frankenstein.

Given the long evolution of book literacy, what features best describe its patterns of growth? For one thing, literacy definitions over time have become more complex—a pattern already reflected in computer literacy. For example, the simpler concepts of operational and instrumental literacy are associated more with the 1970s while the more complex concepts of computer programming and role-related literacy are linked more to the 1980s while the more complex concepts of computer programming and role-related literacy are linked more to the 1980s.

The move from the uncomplicated to the more complex can be seen as follows: operational literacy requires minimal machine skills; instrumental literacy requires minimal machine skills; instrumental literacy, the skills to learn from programmes developed by others;

algorithmic reasoning, the skills to create one's own programmes; and role-related literacy, the skills to perform as competent consumers, workers, and citizens in the developing information society. Given the pattern just noted, another question arises: What features best denote literacy's developing complexities? For one thing, more varieties of radiant skill are now assembled under the literacy tent; for another, the functions served by literacy have multiplied.

Why are the meanings of literacy more encompassing today than they were earlier? Today's meanings stem from a multicentury trend toward an ever more varied content disseminated in a widening array of books, monographs, newspapers, magazines, and journals. When Martin Luther translated the Bible into the German language in the sixteenth century, he provided a powerful new stimulus and a popular means for the citizens of his nation to acquire literacy. At that time, however, literacy was by necessity defined largely in relation to one book.

With the spread of the printing press and the growth and popularization of diverse types of knowledge, literacy expanded to include the skills needed to read more diverse materials. In the nineteenth century, when individuals were pressed to understand the world around them by reading the newly invented newspaper, they had to expand their literacy skills. As reading skills have expanded, literacy functions have become more differentiated. "Scientific", "mathematical," "esthetic," "technological," and "language" denote some of these literacy functions. They also symbolize bodies of knowledge that go substantially beyond those in Luther's day or, for that matter, in Horace Mann's time. Significantly, they imply high expectations for schools, and they help to explain the emergence of compulsory education.

The Differences

Computer literacy has already undergone considerable differentiation. Inherent in Previous descriptions of operational, instrumental, programmatic, and role-related concepts of literacy are such functions as word processing, computing; programming, information processing, and participation in teledemocratic arrangements. On the other hand, Marvin and Winther, through a study of computer specialists, have arrived at a different set of functions.

Employing an electronic mail questionnaire sent to users of the Advanced Research Projects Agency Network (ARPANET), a large computer network, they obtained perceptual data from respondents about "the major types of skills and levels of knowledge that organize computer users into major classes". After analyzing their data, Marvin and Winther posited five levels of users. These classes of users, along with pertinent functions and skills, are described here:

1. Many individuals each day come in contact with microwave ovens, candy machines, cash registers, and related devices; since they perform passive rather than active functions, they do not require particular skills.
2. Members of the next group are called "tourists" or casual end-users. Professors, who rely upon experts for data analysis, fall into this category. The skills "tourists" need are very limited.
3. Programmers constitute the next group. They range from those who have written one workable programme to those who have written model programmes. The skills required by programmers, as already noted, are specialized ones.
4. Near the top of the hierarchy is the fourth group, called "hackers," who deal effectively with computer "logiams" and "bottlenecks". Their skills, which they express inventively, are, highly technical.
5. At the top of the pyramid are "wizards". Noted for their capacity to perform "amazing feats," they can effect results seemingly by "passing their hands over circuit boards". Their skills are of the highest order.

Marvin and Winther concluded that "the definition of literacy and its proper objects is at any time not only a standard of skillful performance but also a social sorting procedure for labeling certain tasks and domains of knowledge as worthy of social attention and effort".

If computer users at IBM, the New York City public Schools, and the Pentagon had filled in the electronic questionnaire, their lists of functions and skills would likely have differed from one another and from those identified by ARPANET users. Given the diversity of their missions and "social sorting" procedures, they would, in other words, have arrived at different skills. Obviously, schools cannot prepare individuals to perform all computer functions; rather, they must concentrate on basic skills relevant to many differentiated functions.

If computer literacy skills are arrived at through processes of social construction which go on in highly specialized organizations such as ARPANET, the following question becomes pertinent: How are current definitions of literacy, as described above, likely to be reconstructed in the decades ahead? The expanded and interrelated uses of micro-technologies provided one context for examining this question. Since the microcomputer is the most visible of all microelectronic technologies, thinkers have understandably concentrated on computer literacy during the last decade.

This approach to literacy has provided a useful focus for designing new learning experiences. However, "computer," as the modifier of literacy, may prove to be an impermanent concept. To be sure, microcomputers are pivotal points in the processing of information; however, they are only part of a much larger and more extended communication complex. The microcomputer is, to use concepts from gestalt psychology, a small "figure" against a more encompassing and expanding "ground".

THE PROGRAMMING

Importance of C in Programming: C is a general-purpose, structured language. Its instructions consists of terms that represent its algebraic expressions. C contains many additional features that allow it to be used at a lower level, thus bridging the gap between machine language and the more conventional high-level languages (HLL). C has got a relatively small instruction set, though actual inplementations include extensive library functions, which enhance the basic instruction.

Like any other programming language, C is a versatile language used for different programmes.

Data Used in Programming: C supports several different types of data, each of which may be represented differently within the computer's memory. The C language contains four basic data types: int, float, char and double. The following table shows different data types and their typical memory requirements.

Data Type	***Description***	***Typical Memory Requirements***
int	Integer quantity	2 bytes or one word (varies from compiler to compiler)
char	Single character	1 byte
float	Floating-point number (number with a decimal point and/or an exponent)	1 WORD (4 bytes)
double	Double-precision floating number (more significant figures and an exponent which may be larger in magnitude)	2 words (8 bytes)

'C' and its Indentifiers: Identifiers are names given to various programme elements, such as variables, functions and arrays. These consist of letters and digits, in any order, except that the first character must be a letter. The features of identifiers include

(a) Both upper and lowercase letters are permitted.

(b) Upper and lower case letters are not interchangeable, *i.e.*, an uppercase letter is not equivalent to the corresponding lowercase letter.

(c) Underscore (_) can be included as a character and is considered to be a letter. Underscore is often used in the middle of an identifier.

Examples

Valid identifiers	Non-valid identifiers
X	4th
Y 12	"X"
sum _1	order-no
TABLE	error flag
names	
tax_rate	

C and the Controlling Statements: C programming language utilizes a number of control statements for controlling proper execution of a programme. This programme requires that a logical test be carried out at some particular point within the programme.

An action can then be carried out whose exact nature depends on the outcome of the logical test. This is called as conditional execution. Finally, there is a special kind of conditional execution in which one group of statements is selected from several available groups. This is called selection.

VARIOUS CONTROLLING ELEMENTS

While Statement

The 'while' statement is used to carry out looping operations. The general form of the statement is:

while (expression) statement

Do-while Statement

The 'do-while' statement enables to have a loop with the test for continuation at the end of each pass. The general form of the statement is:

do statement while (expression);

The included statement will be executed repeatedly as long as the value of expression is not zero.

For Statement

The 'for' statement is the most commonly used control statement in-C. This includes an expression that specifies an initial value for an index, another expression that determines whether or not the loop is continued and a third expression that allows the index to be modified at the end of each pass. The general form of the 'for' statement is:

for (expression 1; expression 2; expression 3) statement

If-else Statement

→ The 'if-else' statement is used to carry out a logical test and take one of the two possible actions depending on the outcome of the test (*i.e.,* whether the outcome is true or false).

The general form of the statement is:

if (expression) statement

The expression must be placed in parentheses.

Switch Statement

→ The 'switch' statement causes a particular group of statements to be chosen from several available groups. The selection is based on the current value of an expression that is included within the 'switch' statement.

The general form of the statement is:

switch (expression) statement

Here, the expression results in an integer value.

Break Statement

→ The 'break' statement is used to terminate loops or to exit from a switch. It can be used within a 'while', a 'do-while' a 'for' or a 'switch' statement.

The general form of the statement is:

break;

This is without any embedded expression or statement.

Continue Statement

→ The 'continue' statement is used to bypass the remainder of the current pass through a loop. The loop does not terminate when a continue statement is encountered. The continue statement can be included within a 'while', a 'do-while' or a 'for' statement.

The general form of the statement is:

continue;

This is without any embedded statements or expressions.

Go to Statement

→ The 'go to' statement is used to alter the normal sequence of programme execution by transferring control to some other part of the programme. The general form of the 'go to' statement is:

go to label;

Here, 'label' is an identifier used to label the target statement to which control will be transferred.

Various Types of Data: The C language deals with a number of structured data types. Some of the most important data structures in C are:

1. Arrays
2. Strings
3. Union
4. Pointer

Role of Arrays: An array is an identifier that refers to a collection of data items, which all will have the same name. The simplest form of an array is a one-dimensional array that may be defined abstractly as a finite-ordered set of homogenous elements.

Arrays are of the following types:

1. One-dimensional arrays
2. Two-dimensional arrays
3. Multi-dimensional arrays

One-dimensional Arrays

→ A one-dimensional array is used when it is necessary to keep a large number of items in memory and reference all the items in a uniform manner.

The C declaration

int b [100];

reserves 100 successive memory allocations, each large enough to contain a single integer.

Two-dimensional Arrays

→ A two-dimensional array is a logical data structure that is useful in programming and problem solving. Here, the component type of an array can be another array.

For example

int a [3] [5];

This defines a new array containing three elements. A two-dimensional array clearly illustrates the differences between a logical and a physical view of data.

Multi-dimensional Arrays

→ An array with more than two dimensions is a multi-dimensional array. Such an array may be declared by

int b[3][2][4];

Similarly, a six-dimensional arra can be defined, for example, as

int C[7][15][3][5][8][2];

Role of String: A 'string' in C language in defined as an array of characters. Each string is terminated by the NULL character, which indicates the end of the string. A string constant is denoted by any set of characters included in double-quote marks. The NULL charac ers is denoted by any set of characters included in a string constant when they are stored. Within a programme, the NULL character is denoted by the escape sequence\0. Other escape sequences that can be used are:

* \n for a new line character
* \t for a tab character
* \b for a backspace character
* \" for the double-quote character
* \\ for the backslash character
* \ ' for the single-quote character

Role of Unions: In C language 'union' is a type of data structure, which allows a variable to be interpreted in different ways. Union basically has two parts: a fixed part and a variable part. The fixed part consists of all member declaration upto the keyword union, whereas the variable part consists of the remainder of the definition.

A union provides different road maps for the same area of memory and allows the programmer to determine the currently required road-map.

Role of Pointers: 'Pointers' in C language allow the programmer to reference the location of objects as well as the objects themselves. For example:

if x is declared as an integer, &x refers that has been set aside to contain x. Here &x is called a pointer.

Again, for example, the declarations

```
int *pi;
float *pf;
char *pc;
```

declare three pointer variables.

'pi' is a pointer to an integer, 'pf is a pointer to a float and 'pc' is a pointer to character.

Pointer values can be assigned like any other values, *e.g.*, the statement pi = &x; assigns a pointer to the integer x to the pointer variable pi.

Pointer could be of the following types: 1. Dangling pointer-contains address of a variable that can be freed. 2. Data type pointer-points to any basic data type. 3. External pointer-points not located in an element list node. 4. Internal pointer-points located inside element list nodes. 5. Null pointer-signals the end of a list.

Theories of OOPS: OOPS is known as object-oriented programming. This programming concept integrates the use of class and objects.

Concept of Class: A 'class' embodies the concept of an abstract data type by defining both the set of values of a given type and the set of operations that can be performed on those values. Object-oriented programming constructs modelled out of data types called classes. A class encloses both the data and the functions that operate on the data, into a single, unit as shown:

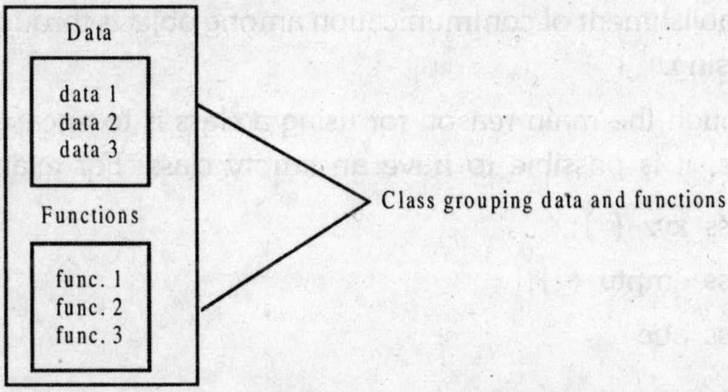

Classes are the basic language construct of C++ for creating the user-defined data types. A class follows the principle that the information about a module should be private to the module unless it is specially declared public.

The syntax of class specification is:

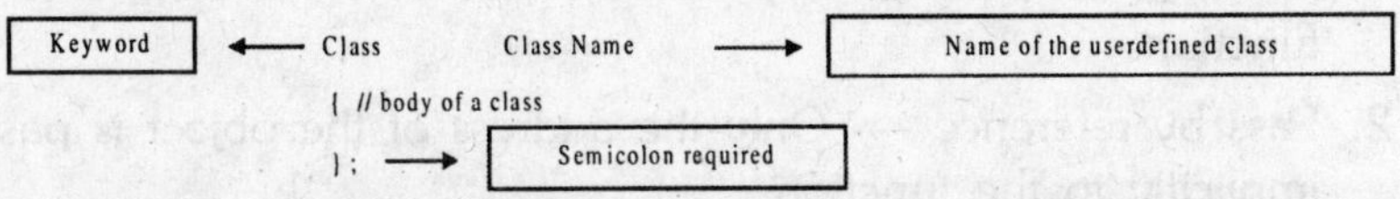

However, more than one class with the same class name in a programme is an error.

What is Object?: 'Object' is an instance of a class. An object is a conceptual entity possessing the following properties:

1. It is identifiable.
2. It has features that span a local space.
3. It has operation that can change the status of the system locally, while also including operations in peer object.
4. It refers to a thing, either a tangible or a mental construct, which is identifiable by the users of the target system.

A class specification only declares the structure of objects. The process of creating objects of the class is called INSTANTIATION.

In OOPs, the process of programming involves the following steps:

1. Creation of classes for defining objects and their behaviours.
2. Creation of class objects; class declaration acts like a blueprint for which physical resources are not allocated.

3. Establishment of communication among objects through message passing.

Although the main reason for using a class is to encapsulate data and code, it is possible to have an empty class. For example:

```
Class xyz { };
Class empty { };
Class abc
{
};
```

are all 'empty' classes.

Like any other data type, an object can be passed as an argument to a function in the following ways:

1. Pass by value → A copy of the entire object is passed to the function.
2. Pass by reference → Only the address of the object is passed implicitly to the function.
3. Pass by pointer → The address of the object is passed explicitly to the function.

Object-oriented programming is based on three major concepts:

* Inheritance
* Polymorphism
* Overloading

Concept of Inheritance

The technique of building new classes from the existing class is called INHERITANCE. It is also a technique of organizing information in a hierarchical form. It allows new classes to be built from older and less specialized classes. The new classes are called derived classes and the existing class are called lease classes.

Types of Inheritance

Inheritance is classified into the following forms based on the levels of inheritance and interrelation among the classes involved in inheritance process:

Single Inheritance: Derivation of a class from only one base class.

Multiple Inheritance: Derivation of a class from several (two or more) base classes.

Hierarchical Inheritance: Derivation of several classes from a single base class.

Multi-level Inheritance: Derivation of a class from another derived class.

Hybrid Inheritance: Derivation of a class involving more than one form of inheritance.

Multipath Inheritance: Derivation of a class from other derived classes.

What is Polymorphism?: Polymorphism follows the concept of allowing a single name to be used for more than one related purpose. The different ways of achieving polymorphism in C++ programme are: 1. Function name overloading; 2. Operator overloading; 3. Dynamic binding.

The types of polymorphism are shown in the following figure:

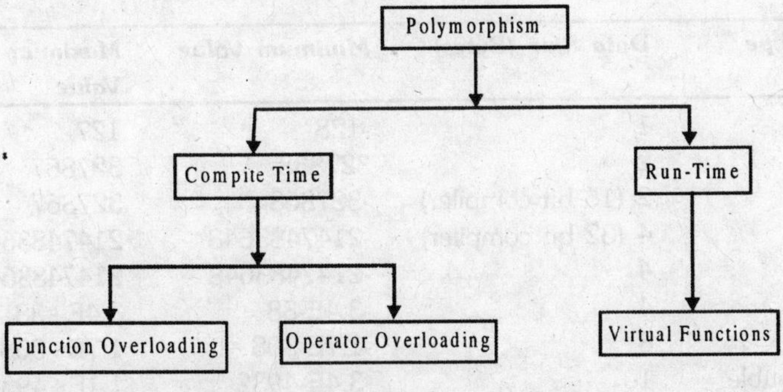

Fig. Polymorphism hierarchy

= Function overloading is realized by invoking a suitable function whose signature matches with the arguments specified in the function call statement.

= Operater overloading is realized by allowing operators to operate on the user-defined data types with the same interface as that of the standard data types.

Kinds of Data in C++: C++ supports the following classes of data types

* Primary (fundamental data types)
* Derived data types
* User-defined data types

C++ language supports the following basic data types

char	a single byte that can hold one character
int	an integer
float	a single precision floating point number
double	a double precision floating point number

Further, applying qualifiers to the above basic types yields additional types. A qualifier alters the characteristics such as the size or sign of the data types. The qualifiers that alter the size are short and long.

These qualifiers are applicable to integers and yield two more types

short int	integer represented by 16 bits irrespective of machine type
long int	integer represented 32 bits irrespective of machine type

The exact sizes of these data types depend on the compiler as shown:

Data Type	***Data Size (Bytes)***	***Minimum Value***	***Maximum Value***
char	1	-128	127
short	2	327838	327867
int	2 (16 bit compiler)	-327868	327867
	4 (32 bit compiler)	-2147483648	2147483647
long	4	-2147483648	2147483647
float	4	-3.4E-38	3.4E+38
double	8	-1.7E-308	1.7E+308
long double	10	-3.4E-4932	1.1E+4932

The qualifier 'long' can also be used along with the double precision floating point type long double → an extended precision floating point number.

The sign qualifiers are signed and unsigned. The sign qualifiers are applicable to the integer data types (int, short int and long int) resulting in six additional data types given below:

= signed short int

= unsigned short int

= signed int

= unsigned int

= signed long int

= unsigned long int

Use of Variables in C++: A variable must be defined before using it in a programme. It reserves memory required for data storage and associates it with a symbolic name. The variable name can be any valid C++ identifier except the reserved words. The data type can be any primitive or user-defined data type such as int, float, double and so on.

The syntax of variable definition is shown below:

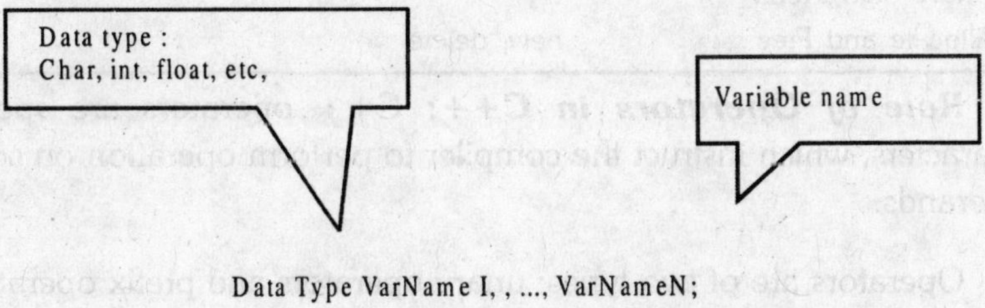

The following are some of the valid variable definition statements

```
int a;                      // a is an integer variable
int b, c, d;                // b, c, and d are valid
                            integer variables
float total, length;        // total and length are
                            valid real variables
double suml, product;       // suml and product are
                            valid double variables
```

Variables can be defined at the point of their usage as follows:

```
int c;
count << "Hello World";
int d = 10;
for (int i = 0; i < 10; i++)
count << i;
```

Note that the variables d and i are defined at the point of their usage.

In C++, the operators as follows are:

Operator Category	***Operators***
Arithmetic	+, -, *, /,
Bit-wise	&, |, ~, ^
Logical	&&, ||, !
Relational	>, <, = =, !=, <=, >=
Assignment or Initialization	=
Arithmetic Assignment	+=, -=, *=, / =, %=, & =, |, ^ =
Shift	<<, >>, <<=, >>=
Unary	+ +, — —
Subscripting	[ ]
Function Call	( )
Dereferencing	- >
Unary Sign Prefix	+, -
Allocate and Free	new, delete

Role of Operators in C++: C++ operators are special characters, which instruct the compiler to perform operation on some operands.

Operators are of two types: unary operators and prefix operators. Unary operators operate on a single operand and prefix operators are indicated before operands.

Importance of Constants

A 'constant' does not change its value during the entire execution of the programme.

Constants can be classified as:

1. Integer constants: (a) Octal (Base 8)
   (b) Decimal (Base 10)
   (c) Hexadecimal (Base 16)
2. Floating point constants: (a) Decimal notation
   (b) Exponential notation
3. Character constants: Enclosed within single quotes
4. String literal constants: Enclosed in double quotes

VARIOUS ELEMENTS

C++ programme consists of many elements, which are identified by the compiler as tokens. Tokens supported in C++ can be categorized as keywords, variables, constants, special characters and operators as shown:

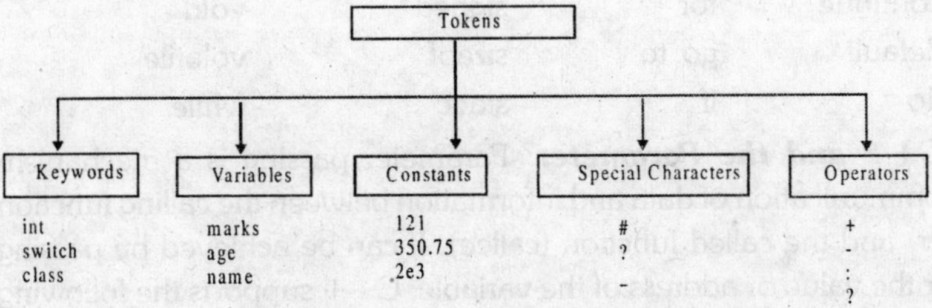

In a C++ programme, every word can be classified as either an indentifier or a keyword. As the name suggests, identifiers are used to identify or name variables, symbolic constants, functions and so on. Keywords have predefined meanings and cannot be changed by the user.

The following rules need to be followed while naming identifiers

- Identifier name is formed by using alphabets, digits or underscore characters.
- Identifier names must begin with an alphabets, digits or underscore character.
- The maximum number of characters used informing an identifier must not exceed 31 characters. Some compilers allow the identifier length to be more than 31 characters however, only the first 31 characters are significant.
- C++ is case sensitive (since the upper and lowercase letters are treated differently).
- C++ has standard identifiers called keywords. Keywords are declared by the C++ language and have a predefined meaning. Hence, they cannot be used for any other purpose other than that specified by the C++ language. The keywords common to C and C++ are:

auto	double	int	struct
break	else	long	switch
case	enum	register	typedef
char	extern	return	union
const	float	short	unsigned
continue	for	signed	void
default	go to	sizeof	volatile
do	if	static	while

C++ and the Parameter: Parameter passing is a mechanism for communication of data and information between the calling function (caller) and the called function (callee). It can be achieved by passing either the value or address of the variable. C++ supports the following three types of parameter passing schemes

- Pass by Value
- Pass by Address
- Pass by Reference (only in C++)

Benefits of Passing by Value

The default mechanism of parameter passing is called pass by value. Pass-by-value mechanism does not change the contents of the argument variable in the calling function (caller), even if they are changed in the called function (callee), because the content of the actual parameter in a caller is copied to the formal parameter in the callee. The formal parameter is stored in the local data area of the callee.

Benefits of Passing by Address

C++ provides another means of passing values to a function known as pass by address. Instead of passing the value, the address of the variable is passed. In the function, the address of the argument is copied into a memory location instead of the value. The de-referencing operator is used to access the variable in the called function.

Passing by Reference

Passing parameters by reference has the functionality of pass by pointer and the syntax of callby-value mechanism. Any modifications made through the formal pointer parameter is also reflected in the actual parameter. Therefore, the function body and the call to it is identical to that of call-by-value but has the effect of call-by-pointer mechanism.

To pass an argument by reference; the function call is similar to that of call by value. In the function declarator, those parameters, which are to be received by reference must be preceded by the operator. The reference type formal parameters are accessed in the same way as normal value parameters.

ROLE OF CONSTRUCTORS

A constructor is a special method of a class that is invoked whenever an object of that class is created. A constructor is always named with the same name as the class itself. The constructor of a class is the first member function to be executed automatically when an object of a class is created.

A constructor has the following properties:

— has the same name as that of the class to which it belongs.

— is executed automatically whenever the class is instantiated.

— does not have any return type.

— normally used to initialize the data members of a class.

— used to allocate resources such as memory, to the dynamic data members of a class.

Syntax

```
class           Class Name
{               .... // private members
                public: → must be public
                // public members
                class name (); → constructor prototype
}
class name      :: class name () → constructor definition
                {
                // constructor body definition
                }
```

Role of Destructors

A class can have a member function called 'destructor', which is invoked when an object is destroyed. This function complements the operation performed by any of the constructors.

For objects that are local non-static variable, the destructor is called when the function in which the object is defined is about to terminate.

For static or global variable, the destructor is called before the programme terminates. Even when a programme is interrupted using an exit () call, the destructors are called for all objects that exist at that time. Destructors have no return type.

Importance of Templates

A template in C++ allows the construction of a family of template functions and classes to perform the same operation on different data types. The templates declared for functions are called function templates and those declared for classes are called class templates.

Syntax of a function template:

How to Handle Exceptions?: Exception handling provides a way of transferring control and information to an unspecified caller that has expressed willingness to handle exceptions of a given type.

9

THE NETWORKING

There are three different types of network for different purposes:

(1) LAN

(2) MAN

(3) WAN

Advantages of LAN

Characteristics

(i) A diameter not more than a few kms.

(ii) A total data rate of at least several Mbps.

(iii) Complete ownership by a single organization.

Use of MAN

Characteristics

(i) Covers entire city, but uses LAN technology.

(ii) Have data rates below 1 Mbps.

(iii) Owned by multiple organizations.

Where to Use WAN

Characteristics

(i) More scalable than LAN and MAN.

(ii) Connects many sites spread across large geographic distances with many computers at each site.

(iii) Provides sufficient capacity to permit the computers to communicate simultaneously.

Importance of Static Allocation

The traditional way of allocating a single channel, such as a telephone trunk, among multiple competing users is FDM (Frequency—Division Multiplexing). If there are N users, the bandwidth is divided into N equal sized portions each user being assigned one portion. Since each user has his own private frequency band, there is no interference between users.

When the number of stations is large and continuously varying, or the traffic is bursty, FDM presents some problems. If the spectrum is cut up N regions and fewer than N users are currently interested in communicating, a large piece of valuable spectrum will be wasted. If more than N users want to communicate, some of them will be denied permission, for lack of bandwidth, even if some of the users who have been assigned a frequency band hardly ever transmit or receive anything.

The poor performance of static FDM can easily be seen from a simply queueing theory calculation. Let us start with the mean time delay, T, for channel of capacity C bps, with an arrival rate of l, frames/sec, each frame having a length drawn from an exponential probability density function with mean 1/m bits/frame,

Divide the single channel up into N independent subchannels, each with capacity C/N bps. The mean input rate on each of the subchannels will now be k/N. Recomputing T we get,

The mean delay using FDM is N times worse than if all the frames were somehow magically orderly arranged in a big central queue.

Role of Dynamic Channel Allocation

Underlying all the work done in this area are five key assumptions.

Station Model: Consists of N independent stations (computers or terminals), each with a programme or user that generates frames for transmission. The probability of a frame being generated in an interval of length At is XAt, where X, is a constant (the arrival rate of new frames).

Single Channel Assumption: Available for all communication. All stations can transmit on it and all can receive from it.

Collision Assumption: If two frames are transmitted simultaneously, they overlap in time and the resulting signal is garbled, called a collision. A collided frame must be transmitted again later. There are no errors other than those granted by collisions.

Continuous Time: Frame transmission can begin at any instant.

Slotted Time: Time is divided into discrete intervals (slots). Frame transmissions always begin at the start of a slot. A slot may contain 0, 1 or more frames, corresponding to an idle slot a successful transmission or a collision respectively.

Carrier Sense: Stations can tell if the channel is in use before trying to use it. If the channel is sensed as busy, no station will attempt to use it until it goes idle.

No Carrier Sense: Stations cannot sense the channel before trying to use it and go ahead and transmit.

Importance of OSI Model: The OSI model is based on a proposal development by the International Standards Organization (ISO) as a first step toward international standardization of the various protocols (Day and Zimmermann, 1983). The model is called the ISO OSI (Open System Interconnection) Reference Model because it deals with connecting open systems that is, systems that are open for communication with other systems.

The OSI model has seven layers. The principles that were applied to arrive at the seven layers are:

1. A layer should be created where a different level of abstraction is needed.
2. Each layer should perform a well-defined function.
3. The function of each layer should be chosen with an eye toward defining internationally standardized protocols.
4. The layer boundaries should be chosen to minimize the information flow across the interfaces.

5. The number of layers should be large enough that distinct functions need not to thrown together in the same layer out of necessity and small enough that the architecture does not become unwieldy.

The OSI model itself is not a network architecture because it does not specify the exact services and protocols to be used in each layer. However, ISO has also produced standard for all the layers.

Role of ISDN Architecture: ISDN data transmission services will allow users to connect their ISDN terminal or computer to any other one in the world. At present such connections are frequently impossible internationally due to incompatible national telephone system.

The key idea behind ISDN is that of the digital bit pipe, a conceptual pipe between the customer and the carrier through which bits flow. Bits can flow through the pipe in both directions.

The digital bit pipe can and normally does, support multiple independent channels by time division multiplexing of the bit stream. The exact format of the bit stream and its multiplexing is a carefully defined part of the interface specification for the digital bit pipe. Two principal standards for the bit pipe have been developed, a low bandwidth standard for home use and a higher bandwidth standard for business use that support multiple channels that are identical to the home use channel. Furthermore, business may have multiple bit pipes if they need additional capacity beyond what the standard business pipe can provide.

An ISDN PBX (Private Branch Exchange) can directly interface to ISDN terminals and telephones. To accommodate these, the customer can install one or more terminal adapters that speak RS-223-C to the terminal and ISDN to the PBX. ISDN PBX depicts five kinds of devices on the customer premises

1. NT1: network boundary
2. NT2: customer PBX
3. TE1: ISDN terminal
4. TE2: non-ISDN terminal
5. TA: terminal adapter

Importance of ISDN Interface: The goal of ISDN is to present the user with a digital bit pipe at either the T or S reference point. The ISDN physical layer deals with the mechanical, electrical, functional and procedural aspects of the interface. ISDN uses a new kind of connector, completely unrelated to the 25-37-and 9-pin D connectors used for RS-232-C and RS-449. The ISDN connector has eight contracts. Two are used for transmit and transmit ground. Two more are used for receive and receive ground. The remaining four are used to allow NT1 or NT2 to power the terminal or vice versa. By using this balanced transmission scheme, like RS-422-A, the ISDN cable can be 1 km long with good noise immunity. The ISDN bit pipe supports multiple channels interleaved by time division multiplexing. Several channel types have been standardized:

A — 4 kHz analog telephone channel

B — 64 kbps digital PCM channel for voice or data

C — 8 or 16 kbps digital channel

D — 16 or 64 kbps digital channel for out-of-band signaling

E — 64 kbps digital channel for internal ISDN signaling

H — 384, 1536 or 1920 kbps digital channel

The ISDN bit stream can be used to support either circuit switching or packet switching depending on how bursty the traffic is.

In the packet-switching scenario, the ISDN customer calls up a nearby IMP. This connection is used to transmit packets from the customer's equipment to the IMP, which transmits them to the final destination via a traditional packet switching network. The advantage of this scheme is that the call to the IMP will generally be a local call, so the charge for the service will be the cost of a local call, plus a certain amount per packet.

Scenario of ISDN: ISDN will provide a physical layer bit stream from homes and offices onto which layers 2 through 7 can built.

There is a world wide agreement on the interface standard for the basis rate and there are only two minor variants for the primary rate. This standardization will lead to large user demand for ISDN equipment, thus leading to mass production, economies of scale and inexpensive

NLSI ISDN chips. Using predictive encoding and other techniques, it is now possible to transmit voice in a 32 kbps channel. By the time ISDN exchanges have been built and installed 16 kbps will be enough. Using the agreed upon 64-kbps channels will thus effectively waste 75 percent of the world's long-distance transmision capacity, a scarce, valuable and expensive resources. One of the key home services predicted for ISDN is access to remote data bases. For data basis that are relatively stable, ISDN may find itself in competition with video disks, with consumers simply buying the entire database on disk for use with their personal computers.

Use of ATM: A significant wide area technology is known as Asynchronous Transfer Mode (ATM). ATM is an attempt to design a single technology that can be used to provide voice, video and data services across a wide area.

Packet size is a key issue in designing ATM, a general purpose network technology. Data achieves maximal throughout when packet sizes are large because the overhead from headers is minimized by carrying a maximum payload with each header. Thus, networks designed to optimize data transport often have packet sizes of 8 Kbytes a larger.

To achieve high data rates, low delay and low jitter, ATM technology divided all data into small, fixed-size packets called cells. Each ATM cell contains exactly 53-octet: 5 octets of header information and 48 octets of data. Furthermore, ATM allows one to specify the quality of service requirements foreach communication. Thus, when using ATM for audio, one must specify both low delay and low jetter. In ATM terminology, a Constant Bit Rate (CBR) is specified for voice or video. In contrast applications that use ATM to send data can request an Available Bit Rate (ABR) connection.

Difference between Subnets Using Datagrams Internally and Subnets Using Virtual Circuits Internally

Issue	***Datagram Subnet***	***VC Subnet***
Circuit setup	Not possible	Required
Addressing	Each packet contains the full source and destination address	Each packet contains a short vc number
Effect of node failures	None, except for packets lost during	All vcs that passed through the failed equipment are

	the crash	terminated
Congestion control	Difficult	Easy if enough buffers can be allocated in advance for each vc set up
State information	Subnet does not hold state information	Each established vc requires subnet table space
Routing	Each packet is routed independently	Route chosen when vc is setup; all packets follow this route
Complexity	In the transport layer	In the network layer
Suited for	Connection-oriented and connectionless service	Connection-oriented service

Role of Virtual Circuits

1. Virtual circuits allow packet to contain circuit numbers instead of full destination addresses. If the packets tend to be fairly short, a full destination address in every packet may represent a significant amount of overhead, and hence wasted bandwidth virtual circuits internal to the subnet has an advantage when many of the "hosts" are actually interactive terminals with only a few characters per packet.
2. Virtual circuits also have vulnerability problem. If an IMP crashes and loses its memory, even if it comes back up a second later, all the virtual circuits passing through it will have to be aborted. In contrast, if a datagram IMP goes down, only those users whose packets were queued up in the IMP at the time will suffer and may be not even all those, depending upon whether they have already been acknowledged or not. The loss of a communication line is fatal to virtual circuits using it, but can be easily compensated for if datagrams are used. Datagrams also allow the IMPs to balance the traffic throughout the subnet, since routes can be changed halfway through a connection.

Importance of Routing

The real function of the network layer is routing packets from the source machine to the destination machine.

The routing algorithm is that part of the network layer software responsible for deciding which output line an incoming packet should be transmitted on. If the subnet uses datagram internally, this decision must be made a new for every arriving data packet.

Stability is also an important goal for the routing algorithm.

Routing algorithms are of two types:

(a) Non-adaptive—These algorithms do not base their routing decisions on measurement or estimates of the current traffic and topology.

(b) Adaptive—These algorithms in contrast follow the above concept.

Functions of Integrity

The following techniques are used to ensure the integrity of data against accidental damage. Parity bits, checksums and cyclic redundancy checks (CRCs). To use such techniques, a sender computers, a small integer value as a function of the data in a packet. The receiver recomputes the function from the data that arrives and compares the result to the value that the sender computed.

A checksum or CRC cannot absolutely guarantee data integrity for two reasons:

(i) if malfunctioning hardware changes the value of a checksum as well as the value of the data, it is possible for the altered checksum to be valid for the altered data;

(ii) if data changes result from a planned attack, the attacker can create a valid checksum for the altered data.

The methods encode transmitted data with a message authentication code (MAC) that an attacker cannot break or forge. Typical encoding schemes use cryptographic hashing mechanisms. For example, one cryptographic hashing scheme uses a secret key known only to the sender and receiver. Only the receiver can unscramble the data; an attacker, who does not have the secret key, cannot decode the message without introducing an error. Thus, the receiver knows that any message that can be decoded correctly is authentic.

How to Access and Enter?: Many computer systems use a password mechanism to control access to resources. Each user has a password, which is kept secret. When the user needs to access a protected resource, the user is asked to enter the password. A simple password scheme works well for a conventional computer system because the system does not reveal the password to others.

Coding System

To ensure that the content of a message remains confidential despite wiretapping, the message must be encrypted. Encryption scrambles bits of the message in such a way that only the intended recipient can unscramble them. Someone who intercepts a copy of the encrypted message will not be able to extract information.

Essential Coding

In many encryption schemes, the key must be kept secret to avoid compromising security. One particularly interesting encryption technique assigns each user a pair of keys. One of the user's keys, called the private key, is kept secret, while the other, called the public key, is published along with the name of the user, so everyone knows the value of the key. The encryption function has the mathematical property that a message encrypted with the public key cannot be easily decrypted except with the private key and a message encrypted with the private key cannot be decrypted except with the public key.

The relationships between encryption and decryption with the two keys can be expressed mathematically. Let M denote a message, pub-ul denote user l's public key and prv-ul denote user l's private key. Then

M =decrypt (pub-ul, encrypt (prv-ul, M))

and

M =decrypt (prv-ul, encrypt (pub-ul, M))

Revealing a public key is safe because the functions used for encryption and decryption have a one way property.

Public key encryption can be used to guarantee confidentiality. A sender who wishes a message to remain private uses the receiver's public key to encrypt the message. Obtaining a copy of the message as it passes across the network does not enable someone to read the contents because decryption requires the receiver's private key.

Importance of Digital Signatures: To sign a message, the sender encrypts the message using a key known only to the sender. The recipient uses the inverse function to decrypt the message. The recipient knows who has sent the message because only the sender has the key needed to perform the encryption. To ensure that encrypted messages are not copied and resent later, the original message can contain the time and date that the message was created.

Theory of Firewall

A packet filter is often used to protect an organization's computers and networks from unwanted Internet traffic. The figure illustrates that the filter is placed in the router that connects the organization to the rest of the Internet. Illustration of a packet filter used as the primary part of a firewall that protects an organization against umvanted traffic from the Internet.

* A packet filter configured to protect an organization against traffic from the rest of the Internet is called an Internet firewall. An Internet firewall is designed to keep problems in the Internet from spreading to an organization's computers.
* Firewalls are the most important security tool used to handle network connections between two organizations that do not trust each other. By placing a firewall on each external network connection, an organization can define a secure parameter that prevents outsiders from interfering with the organization's computers.
* A firewall can lower the cost of providing security. With a firewall a manager can restrict incoming packets to a small set of computers. In the extreme case, the set can contain a single computer.

Concept of Domain Name

The naming scheme used in the Internet is called the Domain Name System (DNS). Each computer name consists of a sequence of alpha-numeric segments separated by periods. For example, a computer in the Computer Science Department at Kerala University has the domain name

mordred.cs.kerala.edu

and a computer in the College of Engineering at Karnataka University has the domain name:

www.eg.karnataka.edu

Domain names are hierarchical, with the most significant part of the name on the right. The left-most segment of a name (mordred and www in the examples) is the name of an individual computer. Other segments in a domain name identify the group that owns the name.

For example, the segments kerala and karnataka each give the name of a university. The domain name system does specify values for the most significant segment, which is called the top-level of the DNS. The table lists the possible top-level domains.

Domain Name	***Assigned To***
com	Commercial organization
edu	Educational institution
gov	Government organization
mil	Military group
net	Major network support centre
org	Organization other than those above
int	International organization
country code	A country

Role of Geography

The DNS allows organizations to use a geographic registration. For example, the Corporation for National Research Initiatives registered the domain.

cnri.reston.va.us

because the corporation is located in the town of Reston, Virginia in the United States. Thus, names of computers at the corporation end in .us instead of .com.

Once an organization owns a particular domain, the organization can decide whether to introduce additional hierarchical structure. A small corporation can choose no additional hierarchy, while a large organization can choose many levels. For example, if the FCI Corporation is small, it might decide that all names have the form:

computer.fci.com

where computer denotes the name assigned to a single computer.

Different Networking

Computer networks can be categorized along several dimensions and by differing criteria; however, there is no single set of exclusive categories into which all types can be fitted, and networks which are similar in some ways will display contrasting characteristics in others. A network consists essentially of two kinds of elements:

— the network nodes which are the actual computing end systems, for example larger computer systems, personal workstations, terminals, *etc.*;

— the connections between these along which data transmissions travel.

Network nodes can have a greater or lesser degree of autonomy in local processing power and ability to control the communication process. The connections can be implemented in various media, but in some way serve to switch data flows between these network nodes, to pass information from a source to a destination in the most appropriate way. As a first stage, networks can usefully be categorized by the type of switching which is used.

There are three basic types of switching:

— circuit switching,

— message switching;

— packet switching.

Circuit switching occurs when a physical link is made between the two network nodes for the duration of a session, along which the data can flow under the control of the end systems. The connection must be set up before the information exchange can begin, and released once the session is completed; in the interim, the circuit is only available to the two participants even though the channel may have spare capacity which is unused.

Message switching does not establish a physical connection, but relies on the network to deliver a message successfully once sent. The network is used as an intermediary: the source end system delivers the message to the network and then has no further involvement; the network will carry the message to its destination as a discrete process.

The source end system is not aware of the details of this process, and does not work interactively with the destination end system. The network may use a broadcast technique to convey the message; or may route it through a series of intermediate staging posts, each of which receives the message, stores it, and forwards it on to the next step when possible.

Packet switching techniques take a more structured approach, dividing an individual message into separate 'packets', each of which can travel independently through the network. The network is responsible for the safe delivery of these packets, creating them from the message dispatched from the source end system at the point of entry to the network, and re-assembling them into the original message at the point of exit for the destination end system. Since the network is dealing with smaller standard size packets in a predetermined way, a faster transmission rate can be achieved. The network is able to make its own decisions on routing, and can change these dynamically as traffic conditions change, with no apparent difference to the end user.

A second characteristic of network types concerns the geographical extent of coverage, and the size in terms of the total number of network nodes which can be feasibly connected. A general classification is to describe networks as either:

— local area networks, which are limited to a single site or building, interconnecting nodes within a single organization;

— wide area networks, which are not geographically limited in scope, and are capable of national and global-level interconnections.

This distinction arises from the very different technologies from which each of these approaches have grown; however, these categories are beginning to break down, as local area networks become more powerful and can themselves be interlinked to form larger entities. An intermediate category is now emerging, which can potentially provide a link between the first two:

— metropolitan area networks, which exploit the new transmission media such as fibre optic cable, to build city-wide capabilities.

Wide area networks can be usefully seen in two groups, distinguishing public and private systems:

— public telecommunications networks are international in scope, and accessible as a publicly available service;
— private data networks may also be on a national or international scale, but are intended for a distinct user group and not for general public access, although they are likely to make use of services offered by public telecommunications operators.

Each of these network types uses networking technologies which are most appropriate for their purposes. These may be quite unlike each other, employing different techniques, principles and protocols which are not easily compatible. Stemming from these different technologies, there are consequent differences in maximum data transmission rates, number of possible network connections, response times and overloading factors, *etc.*

Maintaining the Network

The degree of control which is placed within the network, and the location of this control also arises from these differences in approach. This may be characterized along a continuum, which offers at the extremes:

— a centralized network, with a single main switch to which all nodes are connected, and through which they communicate with each other;
— a decentralized peer-to-peer network, in which each node possesses the same status as any other, and can communicate with any other of its peers on an equal basis.

For large-and small-scale networks, there are various compromise points along this continuum where they can be placed; an inter-network can also be constructed using various different topological arrangements which are interconnected at specific gateways-links from one type to the other type.

Networking of Terminals

Terminal networks are perhaps the simplest form of computer network, consisting of a number of relatively unintelligent terminals, each of which is linked to the central computer system and not directly to other terminals. These two elements work in a masterslave relationship to each other, in that the central system has much more extensive control of the process than the terminal, even though this may in fact be a more powerful machine, such as a personal computer emulating a terminal. In such centralized systems, the connections between terminals and central computer are arranged in a star topology, with communications routed through the central switch; if this should fail the main communication channel is cut off.

Terminal networks are very commonly used in large organizations that wish to pursue a centralized information policy, holding data and information systems at a single main point and providing widespread access to this. There may be a high level of communication requirements for this arrangement, however, where speed of response is important; for example a lending desk terminal in a branch public library may be required to transmit loan information to a central database and check appropriate records in a very few seconds.

Networking between Micro and Mainframe: The growth in use of desktop and personal computers has had a significant influence on the development of centralized systems, swaying the balance of control more towards the end user. This has led to attempts at creating a greater integration and cohesion between information systems at an organizational level, linking the departmental/office information into the corporate database level.

Terminal emulation packages allow the end user to utilize a familiar desktop PC as a central terminal without the need to switch equipment, and increasingly to use the computing power of the PC to improve screen presentation rather than simply emulating a dumb terminal. This provides the user with a more familiar and 'friendly' workspace environment. Developments in software compatibility are integrating the standard software packages in use on both central and local systems for mainstream applications such as databases.

Importance of LAN: A local area network is a means of linking together a collection of computer-based devices, within a limited geographical area, on a more or less equal basis. This equality makes the local area network (LAN) radically different from a terminal-based network, because each of the computer systems connected is an independent self-governing entity working in a cooperative relationship. Local area networks fall principally into two types:

— proprietary networks which are used to interconnect a number of computers and their peripherals of the same type, for example a network of IBM PC or Macintosh personal computers;
— standards-based networks which can interlink computer systems and devices whatever their type and specification, through the use of common network technology.

For both types of LAN, every device connected to the network must communicate through a uniform hardware interface and software protocol to achieve compatibility. Standardization of the communication process is therefore very important to the concept of local area networks. Since no one method of operation would be suitable for all possible needs, a number of alternative 'standards exist, reflected in the different types of LAN available. A local area network will typically be provided to support:

— resource sharing of expensive hardware resources, for example a colour printer, a high-speed modem, or a CD ROM drive accessible to all nodes on the network;
— resource sharing of software and access to specialized facilities;
— file transfer and access to support the sharing of information between different users, and to provide wider access to central information while ensuring consistency by maintaining a single copy of the data;
— messaging and conferencing to enhance organizational communications and cooperative working.

The most commonly found use of proprietary local area networks is to interconnect IBM-PCs within a single site or building. There are broadly two types of PC-based local area network, distinguished by the operating system by which they are managed.

1. In peer-to-peer PC networks all the network nodes interrelate as equals, able to call on the services of other nodes and to offer their own at all times. This is an intuitively natural way of sharing resources, and is cheap and effective to implement for small workgroups.
2. In server-based PC networks, one or more network nodes act as a provider of services for the other client nodes. Clients can only

call on the services of a server node, which is dedicated to this purpose; there are three main types of server: a file server dispenses data and programme files across the network, either in whole or in part; a print server drives the printing devices and manages the print queues; and a communications server controls external links, such as facsimile machines, or modems.

LAN with a Quality: There are two industry wide standards for more broadly based local area networks, known as Ethernet and Token Ring.

Ethernet is the most widely installed local area network at the current time. This describes a specification for a LAN rather than an actual product; there are many implementations of Ethernet from a range of suppliers.

Ethernet aims to provide a high-speed network which gives all nodes stable and equal access to the network, in such a way as to prevent individual stations interfering with each other; and allowing a transmission to be addressed to individual nodes, a group of nodes, or the whole network community. The intention is that all Ethernet installations are compatible with each other, since they use the same set of protocols. The stations on an Ethernet are each linked directly to a main transmission channel in a bus topology. Stations may not transmit at the same time, but must wait for the channel to be free before sending. Transmissions are broadcast to all nodes on the network, but only utilized by those stations to which the message is addressed.

There is a mechanism for the detection of collisions when stations broadcast together, which requires a station to rebroadcast after a random interval. Thus, all stations are in contention for the channel and cannot be guaranteed a fixed rate of access. This method works well for small to medium size networks, where traffic is in irregular short bursts rather than at regular intervals.

Ethernet therefore suits the typical office environment where user activity follows this pattern. It has a relatively high data rate of up to 10 Megabits per second (Mbps) and is able to support large data file transfers without undue disruption to the user. It does not, however, support voice and video transmission, because the nature of the access medium cannot guarantee the continuity necessary for voice or moving image. There are technical limitations to the size of Ethernet networks, both in length of cabling and in the amount of data traffic which can be supported.

The Token Ring network takes a different approach and as the form chosen by IBM for its own local area network has become more popular. The nodes are connected in a ring configuration, with each node receiving information from its 'upstream' neighbour and re-transmitting it to its nearest 'downstream' neighhour Each node, either originates its own transmission, passes on the data exactly as is received, or removes the data which is addressed to itself.

To control transmission collisions, each node has the opportunity to transmit in turn, governed by the passing of an electronic 'token' around the ring. The Token Ring thus provides for both an egalitarian 'round robin' access, and multiple priority levels for urgent messages. The method is self-governing to quite a high degree, and able to react dynamically to changing conditions.

IEEE Standards

International standards development for local area networks has been coordinated by the Institute of Electrical and Electronic Engineers (IEEE) in the United States. This work is intended to develop a framework within which the major types of LAN can be encompassed rather than to specify one single standard for all requirements.

The standards are known by numerical, rather than product names, and cover both Ethernet and Token Ring style networks.

1. The 802.3 standard defines a broadcast bus-based network, which is effectively the same as the Ethernet type.
2. The 802.5 standard defines a token passing ring network similar to the IBM Token Ring local area network.
3. The 802.4 standard describes a token passing bus network, which combines features of the first two types.
4. The 802.6 standard defines a high-speed Broadband network, better known as a metropolitan area network.

LANs Interconnections

As the local area network grows in size and complexity the amount of traffic increases, overloading the network and degrading response time and reliability. A number of devices have emerged to support both the partitioning of over-large LANs and the desirable interlinking of others.

1. A repeater carries messages between two sections of a single network, allowing it to be extended beyond the maximum cable limit.
2. A bridge is a station which is connected to two networks, through which the two LANs can exchange data frames. The networks must be using the same protocols so that they look like one logical network, but only traffic destined for nodes on the other network will pass over.
3. A router is a device for sending data frames between two dissimilar LANs, since the router can deal with both network types.
4. A gateway links a local area network to a different networking environment altogether, such as a wide area network.

The development of standards for local area networks is continuing with the emergence of newer technology and physical media. In particular, the use of fibre optics is having a substantial impact, potentially opening up the geographic and bandwidth limitations in this field to such an extent that distinctions between local and wide area networks are becoming less meaningful.

Significance of WAN: In the wide area environment, the communications process is seen as a multi-stage process. Networks are constructed from point-to-point links between different nodes on a store and forward basis. Data transmission are passed from one node to the next, *en route,* with each node making its own routing decision based on matching the information which it holds in a routing table with the destination address on the data packet, and in the context of current network traffic conditions.

Control and management of the network is therefore distributed throughout, and not located in any single central point. The network can be extended by adding new connections between nodes and making appropriate updates to routing tables. There are likely to be alternative routes between nodes which can be used at different times. The relaying of data between nodes in this way requires the use of protocols to ensure correct delivery and protect against the corruption of transmissions *en route*.

The protocols normally operate both within the end systems and within the network relay nodes themselves. There are several alternative protocols suites which may be in use; however, all nodes on the network must be running the same protocols to communicate successfully, since this is the glue which binds the disparate systems together.

As an indication, *some common wide area protocols are:*

— the X.25 protocols which have been developed for public telecommunications, and ensure international compatibility in this area; these have also been adopted in some private networks;

— the TCP/IP protocol suite, widely used in private networks and in the academic and research world.

Private Data Networks

Private data networks can be constructed from a number of different elements, as appropriate to requirements and conditions. They, may be national or international in scope, and involve the coordination of several different network operators and administrations, using elements from:

— public telecommunications operators, in the form of leased data lines or use of the public data network;

— private network operators, including commercial long-haul network operators and satellite communications suppliers with independent local or wider scale data lines directly owned.

The Leasing Telecommunication Lines

The leasing of dedicated digital data lines is an alternative to use of the public data network. These are private circuits, provided as a carrier by the telecommunications authority, but under the control of the organization which is renting the line.

Leased lines are point to point connections, permanently linking one site with another, rather than dynamically switchable channels. Since they are leased at a fixed annual rate, the greater the use made of the line the more efficient it becomes.

Multiple rented private circuits are used by large organizations to create private networks, linking remote sites together on a point-to-point basis, or providing a high-speed network backbone for a high volume of traffic between main sites. In the United Kingdom BT provide a range of leased circuits with varying data carrying capacities:

— *Kilostream* and Kilostream N lines provide bulk data transmission at rates from 64 Kbps up to about 1 Megabit per second (Mbps);

— *Megastream* lines provide higher speeds of up to 2 Mbps;

— *Satstream* utilizes satellite capacity to offer point-to-multi-point connections over a wide geographical area.

The use of leased lines is a more secure and flexible option for the larger organization, which is able to deploy the systems and protocols which it finds most suitable, rather than having to conform to those used in public systems.

Telecommunications Networks

There are three types of telecommunications networks presently available for use for computer-based information networking; each of these represents a phase of the developmental efforts to improve services over several decades, using quite different technology and approaches.

Public Switched Telephone Networks

This is an extensive network of cabling reaching into almost all inhabited parts of developed countries, and providing international connections worldwide. This made it attractive as an immediately available option for computer networking in its early stages. The PSTN network is designed for voice communication, and therefore requires the use of an interfacing data communication device, usually a modem, to be used.

The network makes use of circuit switching technology for the same reason, and is a rather noisy and unreliable channel, prone to errors and mis-connection. It is poorly suited to fast and reliable data transfer and depends on the end systems alone to maintain quality through the use of suitable protocols.

The PSTN makes inefficient use of the available carrying capacity because the circuit is idle when neither end system is transmitting but cannot be used by waiting nodes. Since a switched phone line is charged on a time and distance basis, regardless of the amount of traffic, it is not ideal for inter-computer communication over long distances.

Public Data Networks

The public data networks have been constructed specifically for the transmission of computer information, and cannot be used for voice conversations. They provide fast and reliable data transfer, more flexibility and international communications at a reasonable cost, since they are charged for by the amount of data transferred.

The *advantages of packet switching* are the following:

— there is no need for a direct physical connection between sender and receiver;

— information from different calls can then be interleaved on the network channels to make maximum use of the available bandwidth;

— the user, therefore, pays only for information actually transmitted, rather than for elapsed time regardless of the transfer activity;

— standard use of communication protocols facilitates the interconnection of equipment from different manufacturers, operating at different data rates.

The X.25 protocol set is the standard for public packet switched networks. This is an open standard which is available for end systems as well as internally to the network.

Public data networks are mostly managed on a national basis with international connections between them, and so X.25 forms something approaching a global standard in this area. There are, however, two versions of the X.25 protocol; the latest 1988 version is not used by all public telecommunications operators. The public data network is essentially a store and forward network which operates at speeds fast enough to give the impression of a direct connection to the end user.

The network can be considered as an opaque entity—a 'cloud'— by the end systems which need only to specify an address for the remote end system which is being called, having no concern with the details of routing. The network 'cloud' consists of a collection of packet switching exchanges (PSEs) which forward packets through the high-speed network from the local exchange to the destination PSE.

Integrated Services Digital Networks

The development of public data networks for digital communications led to the existence of different networks for different purposes-one for voice, one for data-and requiring organizations using these systems to connect to both. A duplication of networks is undesirable, and also works against the integration of information types, for instance the coordination of voice, data, and image, which is increasingly needed to support multimedia information networking. Two *development programmes in the telecommunications infrastructure* provided the platform for the development of integrated services digital networks:

1. The digitization of the trunk network lines, the main arteries used by the network operator.
2. The replacement of electro-mechanical local telephone exchanges by digital switches, such as System X and System Y.

The ISDN aims to build a public network providing end-to-end digital connectivity which is capable of supporting a range of digital devices and services, both voice and non-voice, on the same basis. This allows users to send and receive data, text and image from computer systems as well as offering an enhanced telephone service. This is achieved over a single ISDN connection, so that one network may be used for all applications, at the same time as providing higher transmission speeds. The ISDN approach is to enable the existing local telephone circuitry-the 'local loop-to carry digital signals through the use of specialized equipment and adapters, rather than a wholesale and long-term replacement programme.

This requires the installation of suitable network connections in the customer premises, which acts as the interface to the digital network. Despite this fully digital transmission system it is still not possible to connect all computer equipment directly to the network, unless the equipment is designed or adapted to work-with the ISDN system. Access to the ISDN network is available at two different capacity rates, known as basic rate and primary rate.

The access rate refers to the number of logical channels which are available on a single ISDN connection. Each of these channels constitutes a circuit which is independently usable by different information media, and can be switched to different destination points.

The Basic Access Rate provides two *bearer* channels and one *data* channel, and for this reason is otherwise known as 213 + D. Each of the bearer channels can simultaneously carry a separate information stream, at a transfer rate of 64 Kbps—for example, one for a voice call and the other for data.

The Primary Access Rate gives a higher number of bearer channels, although the exact number is variable between countries—the United Kingdom, Europe and Australia supply 30 bearer channels to the user in one ISDN connection (known as SOB + D); in the United States, Japan and South East Asia 23 channels are used. SOB + D gives an overall data rate of some 2 Mbps (the equivalent to a BT Megastream leased line) and necessitates the use of extra equipment at the subscriber premises to manage the multiple channels.

Potential benefits from the use of ISDN are the following:

— all 30 ISDN connections can be switched to different locations;

— clearer calls and more reliable data transfers, with less line noise;

— the ability to send two information streams to the same locations, to support remote cooperative working; for example, two users might discuss over the phone the contents of a computer screen which can be transferred between their terminals;

— enabling domestic services, such as home shopping and banking, online database access, security, and control services for the home;

— feasible desktop video-conferencing;

— providing a better infrastructure for tele-working, for example by providing a gateway to an office LAN for home-based workers.

Frame Relay Networks

Frame relay networks technologies are in development which may threaten to overwhelm ISDN before its full potential is reached. These involve the use of high-speed packet switching on high bandwidth networks which will offer much greater information transmission capabilities than the present public data networks. Frame relay networks are made possible by use of less noisy transmission media, such as fibre optic cable, in the network infrastructure, exploiting this greater reliability to simplify the protocols necessary and thus speed up the transmission process.

Integrated Broadband Networks

Integrated broadband networks are sometimes referred to as broadband ISDN (B-ISDN) systems and will also exploit the advantages of fibre optic to offer a dramatically enhanced communication facility, far more powerful than exists at present and able to deliver a wider range of services more effectively.

The intention of developing integrated broadband communications systems is to provide a single network for all services, which can be flexibly and dynamically configured according to individual circumstances and priorities. Digital networks allow both data, voice and image to be sent along the same connection, but these information types have different desirable transmission characteristics.

Data and text require a high degree of accuracy, since the quality of the received information is more easily affected by corruption, but are more tolerant of short transmission delays. Voice and the moving image, on the other hand, are more tolerant of errors since these can be interpreted by human interpretation, but are more susceptible to variations in transmission delays. Existing or Narrowband ISDN circuits do not have a high degree of flexibility in adjusting to these varying demands: for instance, a voice call does not use the full capacity of a 64 Kbps channel, but the spare cannot be freed up for use by others during a session.

Broadband networks aim to overcome such problems by using a universal fast packet switching technique for all information streams, and allocating 'bandwidth on demand' rather than relying on fixed bandwidth channels. Two rates have been proposed for B-ISDN: a basic rate service with a digital access rate of 155 Mbps, and a primary rate service at 622 Mbps, although the full specifications for these have yet to be defined.

A technique known as the Asynchronous Transfer Mode (ATM) is used to transfer the data. This divides the data stream into cells, which are smaller than X.25 packets in size, and can travel through the network at a faster rate. Since the channel is more reliable, quality control is performed at each end of the link and not by each intermediate network switch as well. This ATM approach is particularly suitable for information activity which is inherently 'bursty' in nature.

For example, a service would be able to ensure fast delivery of high definition images, and then free the channel for other use during the reading and thinking time in which the user is not generating any network activity. This will allow the development of true multimedia information services.

Developments in the wider networking field are increasing aiming to deliver faster data transfer rates, on integrated networks which make no distinction between information types.

As the amount of electronic information available across networks grows exponentially, and the sophistication of end-user computing equipment encourages the sharing of information in more intensive forms, network infrastructures respond with the ability to communicate sophisticated forms fast enough to make actual use of the network transparent to the user. At the organizational level, local area networks are also undergoing technical developments which increase their speed and potential for the delivery of more sophisticated and high-powered information applications.

A situation in which these applications can make use of the network 'on demand' in a flexible manner which can respond to changes in circumstance is foreseeable. There is, however, much effort required to make this a reality; the current situation reflects a confusing picture of many alternative networking options which conflict and compete with each other, and which are poorly suited to widespread interconnection between different types of network. The concept of connectivity stresses the importance of this ability to interconnect networks and information networking applications in a reliable and effective manner, and without the need for equipment which is difficult to obtain, install, and use, and soon becomes redundant.

10

System of Operation

Role of Virtual Memory

People were first confronted with programmes that were too big to fit in the available memory. The solution usually adopted was to split the programme into pieces, called overlays. Overlay 0 would start running first. When it was done, it would call another overlay. Some overlay systems were highly complex, allowing multiple overlays in memory at once. The overlays were kept on the disk and swapped in and out of memory, by the operating system, dynamically, as needed.

Although the actual work of swapping overlays in and out was done by the system, the work of splitting the programme into pieces had to be done by the programmer. Splitting up large programmes into small, modular pieces was time consuming and boring.

The method that was devised (Fortheringham, 1961) has come to be known as virtual memory. The basic idea behind virtual memory is that the combined size of the programme, data and stack may exceed the amount of physical memory available for it. The operating system keeps those parts of the programme currently in use in main memory and the rest on the disk. For example, a 16 MB programme can run on a 4 MB machine by choosing which 4 MB to keep in memory at each instant, with pieces of the programme being swapped between disk and memory as needed.

Virtual memory can also work in a multiprogramming system, with bits and pieces of many programmes in memory at once. While a programme is waiting fey-part of itself to be brought in, it is waiting for 1/0 and cannot run, so the CPU can be given to another process.

Importance of Paging

Most virtual memory systems use a technique called paging. On any computer, there exists a set of memory addresses that programmes can produce. When a programme uses an instruction like:

MOV REG, 1000

It does this to copy the contents of memory address 1000 to REG (or vice versa, depending on the computer). Addresses can be generated using indexing, base registers, segment registers and other ways.

* These programme generated addresses are called virtual addresses and form the virtual address space.
* On computers without virtual memory, the virtual address is put directly onto the memory bus and causes the physical memory word with the same address to be read or written.
* When virtual memory is used, the virtual addresses do not go directly to the memory bus. Instead, they go to an MMU (Memory Management Unit) that maps the virtual addresses onto the physical memory addresses.

The virtual address space is divided up into units called pages. The corresponding units in the physical memory are called page frames. The pages and page frames are always the same size. In this example they are 4 KB, but page sizes from 512 bytes to 64 KB have been used in real systems. With 64 KB of virtual address space and 32 KB of physical memory, we get 16 virtual pages and 8 page frames. Transfers between RAM and disk are always in units of a page.

When the programme tries to access address 0, for example, using the instruction

MOV REG, 0

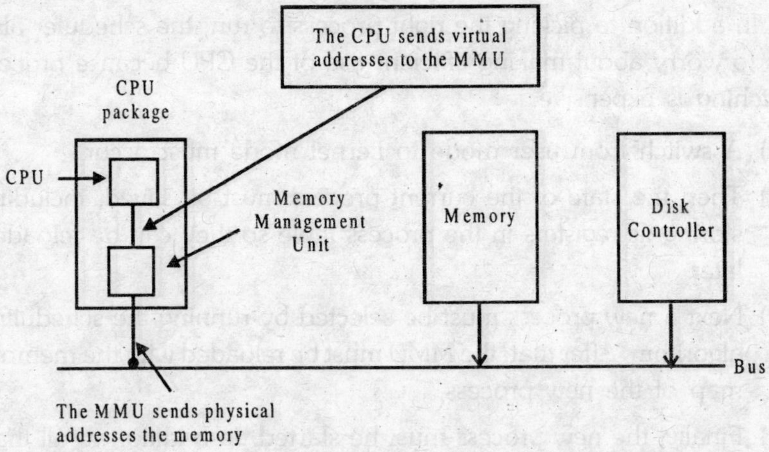

Fig. The position and function of the MMU. Here the MMU is shown as being a part of the CPU chip

TASK OF SCHEDULING

When a computer is multi-programmed, it frequently has multiple processes competing for the CPU at the same time. This situation occurs whenever two or more processes are simultaneously in the ready state. The part of the operating system that makes the choice is called the scheduler and the algorithm it uses is called the scheduling algorithm.

Many of the same issues that apply to process scheduling also apply to thread scheduling, although some are different.

With the advent of personal computers, the situation changed in two ways. First, most of the time there is only one, active process. A user entering a document on a word processor is unlikely to be simultaneously compiling a programme in the background. When the user types a command to the word processor, the scheduler does no have do much work to figure out which process to run the word processor is the only candidate.

When we turn to high end networked workstations and servers, multiple processes often do compete for the CPU, so scheduling matters again. For example, when the CPU has to choose between running a process that updates the screen after a user has closed a window and running a process that sends out queued email, it makes a huge difference in the perceived response.

In addition to picking the right process to run, the scheduler also has to worry about making efficient use of the CPU because process switching is expensive.

(i) A switch from user mode to kernel mode must occur.

(ii) Then the state of the current process must be saved, including storing its registers in the process table so they can be reloaded later.

(iii) Next a new process must be selected by running the scheduling algorithm. After that, the MMU must be reloaded with the memory map of the new process.

(iv) Finally, the new process must be started. In addition to all that, the process switch usually invalidates the entire memory cache, forcing it to be dynamically reloaded from the doing too many process switches per second can chew up a substantial amount of CPU time, so caution is advised.

The Processing

Typically the CPU runs for a while without stopping, then a system call is made to read from a file or write to a file. When the system call completes, the CPU computes again until it needs more data or has to write more data and so on.

Categories of Scheduling algorithms.

Batch: No users impatiently waiting at their terminals for a quick response. Consequently, non-pre-emptive algorithms with long time periods for each process are often acceptable. This approach reduces process switches and thus improves performance.

Interactive: Pre-emption is essential to keep one process from hogging the CPU and denying service to the others. Even if no process intentionally run forever, due to a programme bug, one process might shut out all the others indefinitely. Pre-emption is need to prevent this behaviour.

Real Time: Pre-emption is, oddly enough, sometimes not needed because the processes know that they may not run for long periods of time and usually do their work and block quickly. The difference with interactive systems is that real-time systems run only programmes that are intended to further the application at hand. Interactive systems are general purpose and may run arbitrary programmes that are not cooperative or even malicious.

Different Algorithms

In order to design a scheduling algorithm some goals depend on the environment (batch, interactive or real time), but there are also some that are desirable in all cases.

Under all circumstances, fairness is important. Comparable processes should get comparable service. Giving one process much more CPU time that an equivalent one is not fair. Of course, different categories of processes may be treated very differently. Think of safety control and doing the payroll at a nuclear reactor's computer centre. Somewhat related to fairness is enforcing the system's policies. If the local policy is that safety control processes get to run whenever they want to, even if it means the payroll is 30 sec. late, the scheduler has to make sure this policy is enforced.

Theory UNIX: Although different versions of UNIX differ in subtle ways, the material presented here applies to all of them. The focus here is how UNIX appears at the terminal.

Targets of UNIX: UNIX is an interactive system designed to handle multiple processes and multiple users at the same time. It was designed by programmers, for programmers, to use in an environment in which the majority of the users are relatively sophisticated and are engaged in (often quite complex) software development projects. In many cases, a large number of programmers are actively cooperating to produce a single system, so UNIX has extensive facilities to allow people to work together and share information in controlled ways. The model of a group of experienced programmers working together closely to produce advanced software is obviously very different from the personal computer model of a single beginner working alone with a word processor and this difference is reflected throughout UNIX from start to finish. At the lowest level, a file should just be a collection of bytes. Having different classes of files or sequential access, random access, keyed access, remote access, *etc.* (as mainframes do) just gets in the way. Similarly, if the command.

Is A*

Means list all the files beginning with "A" then the command

Rm A* .

Should mean remove all the files beginning with "A" and not remove the one file whose name consists of an "A" and an asterisk. This characteristic is sometimes called the principle of least surprise. Programmers generally want is power and flexibility. Thus a system should have a small number of basic elements that can be combined in an infinite variety of ways to suit the application. One of the basic guidelines behind UNIX is that every programme should do just one thing and do it well. Thus a compiler do not produce listings, because other programmes can do that better.

Finally, most programmers dislike useless redundancy. To extract all the lines containing the string "ard". from the file f, the UNIX programmer types

Grep and f.

Experiments with UNIX: A UNIX system can be regarded as a kind of pyramid. At the bottom is the hardware, consisting of the CPU, memory, disks, terminals and other devices. Running on the bare hardware is the UNIX operating system. Its function is to control the hardware and provide a system call interface to all the programmes. These system calls allow user programmes to create and manage processes, files and other resources.

The layer in a UNIX system.

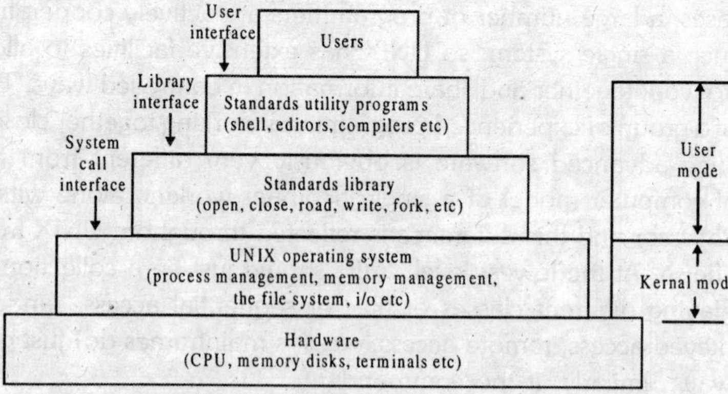

Programmes make system calls by putting the arguments in registers (or sometimes, on the stack), and issuing trap instructions to switch from user mode to kernel mode to start up UNIX. Since there is no way to write a trap instruction in C, a library is provided, with one procedure per system call. Each one first puts its arguments in the proper place, then executes the trap instruction. Thus to execute the read system call, a C programme can call the read library procedure. The library interface is specified by POSIX. POSIX tells which library procedures a conformant system must supply, what their parameters are, what they must do and what results they must return.

In addition to the operating system and system call library, all versions of UNIX supply a large number of standard programmes, some of which are specified by the POSIX 1003.2 standard and some of which differ between UNIX versions. These include the command processor (shell), compilers, editors, text processing programmes and file manipulation utilities.

In the most general case, exec has three parameters: the name of the file to be executed, a pointer to the argument array and a pointer to the environment array.

System Calls Relating to Processes

System Call	***Description***
Pid = fork()	Create a child process identical to the parent
Pid = waitpid (pid & statloc, opts)	Wait for a child to terminate
S = execve (name, argv, envp)	Replace a process' core image
Exit (status)	Terminate process execution and return status
S = sigaction (sig & act & oldact)	Define action to take on signals
S = sigreturn (& context)	Return from a signal
S = sigprocmask (how, & set, & old)	Examine or change the signal mask
S = sigpending (set)	Get the set of blocked signals
S,= sigsuspend (sigmask)	Replace the signal mask and suspend the process

S = kill(pid, sig)	Send a signal to a process
Residual = alarm (seconds)	Set the alarm clock
S = pause ()	Suspend the caller until the next signal

FILING PROCESS IN UNIX: The most visible part of any operating system, including UNIX, is the file system. The UNIX design is especially interesting because it clearly illustrates with minimal mechanism and a very limited number of system calls, a powerful and elegant file system.

Basic Idea

A UNIX file is a sequence of 0 or more bytes containing arbitrary information. No distinction is made between ASCII files, binary files or any other kinds of files. The meaning of the bits in a file is entirely up to the file's owner. The system does not care. File names were originally restricted to 14 arbitrary characters, but Berkeley UNIX increased the limit to 255 characters and this has been adopted by System V and most other versions as well. All the ASCII characters except NUL are allowed in file names, so a file name consisting of three carriage returns is a legal file name.

Files can be grouped together in directories for convenience. Directories are stored as files and to a large extent can be treated like files. Directories can contain subdirectories; leading to a hierarchical file system. The root directory is called/and usually contains several subdirectories. The/character is also used to separate directory names, so that the name/usr/ast/x denotes the file x located in the directory ast, which itself is in the/usr directory.

Some of the Major Directories in UNIX System

Directory	***Contents***
Bin	Binary (executable) programmes
Dev	Special files for I/O devices
Etc.	Miscellaneous system files
Lib	Libraries
Usr	User directories

How to Name a File?: There are two ways to specify file names in UNIX, both to the shell and when opening a file from within a programme. First using an absolute path, which means telling how to get to the file starting at the root directory. An example of an absolute path is/usr/ast/books/most/chap-1. This tells the system to look in the root directory for a directory called use, then look there for another directory, ast. In turn, this directory contains a directory books, which contains the directory.

UNIX allows users and processes to designate the directory in which they are currently working as the working directory. Path names can also be specified relative to the working directory. A path name specified relative to the working directory is a relative path. For example, if/usr/ast/books/most is the working directory, then the shell command:

Cp chap-1 backup1

i.e., has exactly the same effect as the longer command

Cp/usr/ast/books/most/chap-I/usr/ast/books/mos2/backup1

It frequently occurs that a user needs to refer to file that belongs to another user, or at least is located elsewhere in the file tree.

FILES AND CALLS

When a programme starts executing in the standard way, file descriptors 0, 1 and 2 are already opened for standard input standard output and standard error, respectively. In this way, a filter, such as the sort programme, can just read its input from file descriptor 0 and write its output to file descriptor 1, without having to know what files they are. This mechanism works before the programme is started. Some system calls relating to files are given below:

System Call	***Description***
Fd = create(name, mode)	One way to create a new file
Fd = open(file, how...)	Open a file for reading, writing or both
S = closed(fd)	Close an open file
N = read(fd, buffer, nbytes)	Read data from a file into a buffer
N = write(fd, buffer, nbytes)	Write data from a buffer into a file
Position = Iseek(fd, offset, whence)	Move the file pointer

S = stat(name, & buf)	Get a file's status information
S = fstat(fd, & buf)	Get a file's status information
S = pipe(& fd [01)	Create a pipe
S = fcntl(fd, cmd,)	File locking and other operations.

11

SOFTWARE SYSTEM

System software is the collection of system programmes that perform a variety of functions, *viz.* file editing, resource accounting, 10 Management, storage management, *etc.*

SYSTEM FOR MACHINE

A system programme (SP) is a programme which aids in effective execution of a general user's computational requirements on a computer system.

"System programming" is the activity of designing and implementing system programme.

Linker: Modules to each other.

Loader: Rutting the programme inside the memory at a particular location.

The primary goal of the SP designer is to design and code the programme so that the task is done not only correctly but also effectively.

Role of Assembly Language: An assembly language is a lowest level programming language for a computer. It is peculiar for a certain computer system and is hence machine dependent.

FUNDAMENTAL TRAITS

Mnemonic Operation Code: The mnemonic operation codes provide a better understanding of a programme and thus helps in developing or debugging a programme.

Symbolic Operand Specifications: Operand can be specified in the form of symbolic reference rather than their absolute address or value.

Declaration of Data/Storage Area: Data constants can be declared in decimal notations or hexadecimal notations and even storage areas can be defined as a particular size. These data values and storage areas can be referenced by their symbolic names. Statements of an assembly language programme:

Imperative Statements (Machine inst.): Indicates actions to be performed during execution of the assembled programme. Hence each imperative statement exists in the final machine language programme and is translated by the assembler in the format required by the machine.

Declarative Statements (Pseudo op codes): Declares constants or storage areas in a programme.

For example,

The Statement

```
A      DS      1
```

Simply declares a storage area of 1 words known by the name A.

The Statement

```
G      DS      200
```

Indicate that the block of data (200 storage words) is to be reserved. G stands for the first word.

The other words can be accessed as G + 1, G + 2, ... G + 5 is the 6th word in the block.

Constants Declared

The statement

	ONE	DC	'1'
		reserved a location	We can load ONE
ONE			EQV

'I' will not have any reference.

It just replaces the value 1 wherever "ONE" comes.

Assembler Directives (Pseudo op Code): Neither represents machine instruction to be included in the object nor indicates the allocation of storage for constants or programme variables. These statements direct the assemblers to take certain action during the process of assembling the programme. They are also used to indicate certain things regarding how assembly of the input programme to be performed. For example

1. START 100
2. ADDITION START 100

   ADDITION END.

1. Indicates that the first word of the object programme generated by the assembler should be placed in the machine location having address 100.
2. Indicates that no more assembly language statements remain to be processed.

Relocation of Programme

Feasibility of relocating a programme and the manner in which relocation can be carried out characterises a programme into one of the following forms:

(i) Non-relocatable programmes

(ii) Relocatable programmes

(iii) Self-relocating programmes

Non-Relocatable Programmes: A non-relocatable programmes is one which cannot be made to execute in any area of storage other than the one designated for it at the time of its coding or translation. Non-relocatability is the result of address sensitivity of code and lack of information regarding: (i) which parts of the programme are address sensitive and (ii) in what manner.

Relocation is feasible only if relevant information regarding the address-sensitivity of a programme is available. A programme form which does not contain any such information is non-relocatable. According to this definition, a programme phase is a non-relocatable programme form.

Relocatable Programmes: A relocatable programme form is one which consists of a programme and relevant information for its relocation. Using this information, it is possible to relocate the programme to execute from a storage area other than the one designated for it at the time of its coding or translation. The relocation information would identify address-sensitive portions of code and indicate how they can be relocated. For example, in the case of programme A assuming 'absolute address' kind of addressing mode for the instructions, the relocation information can be a table containing the addresses of those instructions which need to be relocated, or it could simply be the address of first and last instruction of the programme, assuming all instructions to be contiguous.

It should be noted that in the relocatable form, the programme is a passive object. Some other programme must operate on it using the relocation information, in order to make it ready-to execute from its load area. The relocatable programme form is called an object module and the agent which perform its relocation is obviously the linkage editor or linking loader. The object module is illustrated below. For the target programme produced by the translator to be relocatable, the translator should itself supply the relocation information along with the programme. Thus, the output interface of every translator (other than a translator which uses the translate-and-go schematic) should be an object module and not merely the code and data constituting the target programme.

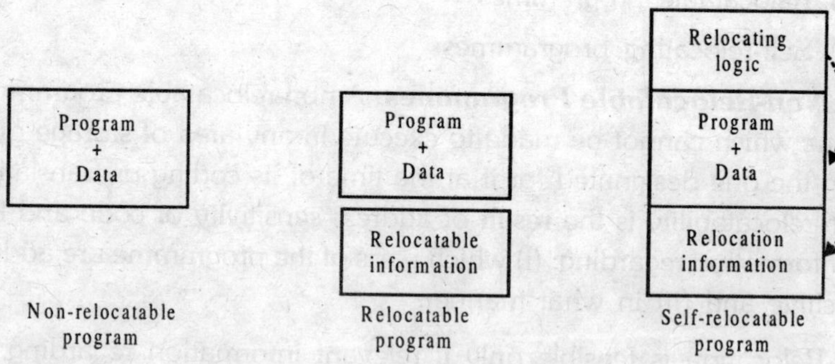

Self-relocating Programmes: A self-relocating programme is one which can itself perform the relocation of its address-sensitive portions. Thus, it not only contains information regarding its address-sensitivity but also the code to make use of this information to relocate the address sensitive part of its code. Such a programme does not need an external agency like the linkage editor to relocate it. It can be simply loaded into its load area and given control for execution. Prior to execution of its main code, the relocating, logic would be executed to adopt the programme to its execution time storage addresses.

Editing the Text

Line Editor: The line editor supports editing operations whose basic unit is a line. The basic limitations of such editors are that part of a line cannot be processed and that the text to be edited has to be explicitly addressed. It incorporates a context search, which permitted the user to indicate a position in the text by specifying a character string. The editor would search for the string and position itself at a line containing the character string. Another facility permitted a relative line specification, viz +nm or -mm, whereby the current position in text would be incremented by nn lines, or decremented by mm lines before performing the next edit operation.

Stream Editor: The stream editor removed all restrictions of line editors. The editor looked upon the entire text as a single continuous stream of characters. Thus, context searches and editing operations could be performed uniformly through the text. One difficulty with stream editors was that the user would prefer to retain the line-oriented structure of the text, *viz.* a programme as a sequence of statement lines, manuscript as a sequence of lines, *etc.*, while the editor would prefer not to have end-of-line characters in the text.

The editor supports character, line and context-oriented commands, using the concept of a pointer into user text (the buffer pointer). The pointer can be manipulated by using some positioning or search commands. TECO also supports advanced features like iterative command sequences and Q-registers which can be manipulated like registers/accumulators in a computer.

Screen Editors: Even though stream editors remove most restrictions of line editors, they too can be quite difficult to use, especially in an online interactive environment.

Working principle of screen editor

The text is conceptually split into screens. At any time, a screen is displayed on the terminal. The user has full control over the terminal and can perform any editing operation he desires. He can move the cursor over the screen, position it at a point where he desires to perform some editing and then go ahead to perform the editing directly. For example, he can overtype an existing string which he wishes to replace. He can bring the cursor over a character to be deleted and press the delete key. The character would be deleted and characters to its right would move left to fill the vacancy. Pressing delete key once again will imply deleting the next character. On insertion, characters following the cursor would move right to the text (*e.g.*, end-of-line characters start-of table indicators), which can also be edited like any other characters in text. The editors might also permit windowing over text so that any arbitrary piece of text could be viewed as a screen for editing purposes. This further extends the editing freedom given to the user.

Some TECO Commands (DEC-10 version)

Command Code	Example	Explanation
C	nC	Moves pointer forward by n characters
R	nR	Moves pointer backward by n characters
J	J	Initialises position of buffer pointer
ZJ	ZJ	Moves pointer to end of buffer
S	nS string	Searches for n t' occurrence of string starting from current pointer position
FS	FS string 1$ string 2$$	Searches for string I and replaces it by string 2
L	±nL	Moves pointer forward/backward by n lines
D	nD	Deletes n characters
I	Istring	Inserts string at current position of pointer

EDITING THE FORMAT

A stream editor helps the search but it is of little use unless a unique string can be indicated. A screen editor permits browsing, but in a strictly sequential manner. Effective support for browsing would require a capability for the user to specify a document structure and then specify his editing requirements in terms of the structure.

NLS (short form for online System) is the earliest structure editor [Engelbart, English 1968]. It is oriented towards document editing. Each document has a hierarchic structure, with different levels like group, plex, branch and statement. Within a section, we can talk of a paragraph, the preceding/succeeding paragraph, first statement in the paragraph, *etc.* Thus, editing can be in terms of structure of document and not merely its content.

Structural awareness by the editor has been found especially useful while editing programme texts since, it eliminates the time consuming process of removing syntax errors. This has led to the development of many structure editors for programme text, the so called syntax directed editors. Here again, the editor gives a conceptually hierarchic structure to the text and permits the programmer to specify structural referables during editing. To further facilitate editing, they permit the programmer to work in terms of syntax elements within a programme statement. Syntax editors are commonly used in programming environments.

Most editors support the following four functions in one form or another—travelling, editing, viewing and display.

Travelling implies movement of the editing context to a new position within the text. This may be done explicitly by the user (*e.g.*, the line number command of a line editor).

Viewing implies formatting the text in a manner desired by the user.

The display component may determine where on the screen a particular view is to appear. The distinction between viewing and display gives rise to interesting possibilities, like multiple views, *i.e.*, windows, appearing on the same screen, multiple independent edit operations using the same display terminal, *etc.*

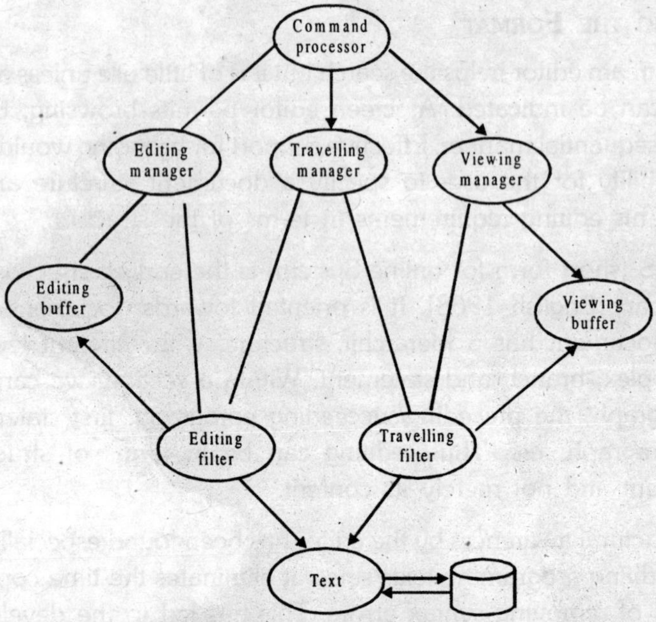

Fig. A schematic of editor structure

An important function of the editor is to maintain compatibility of the two views. This is implemented by using two filters, the edit filter and the viewing filter, both of which operate on the internal form of text.

Importance of Compilers

A compiler is a programme that reads a programme written in one language, the source language and translate it into an equivalent programme in another langauge—the target language. As an important part of this translation process, the compiler reports to its user the presence of errors in the source programme.

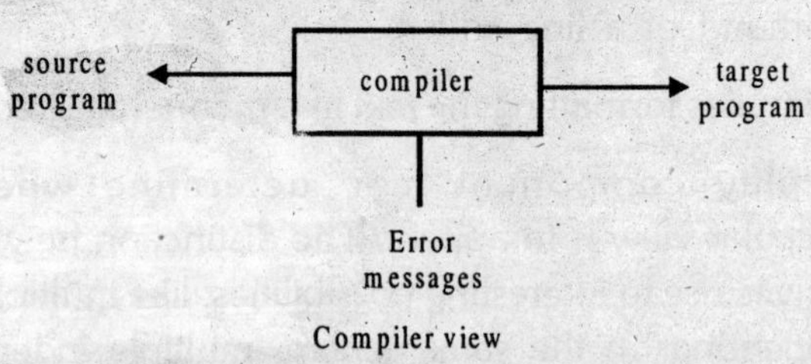

Compiler view

There are thousands of source languages, ranging from traditional programming languages such as Fortran and Pascal to specialized languages that have arisen in virtually every area of computer application. Target languages are equally as varied; a target language may be another programming language, or the machine language of any computer between a microprocessor and a super computer. Compilers are sometimes classified as single-pass, multi-pass, load-and-go, debugging or optimizing, depending on how they have been constructed or on what function they are supposed to perform.

The basic tasks that any compiler must perform are essentially the same. By understanding these tasks, we can construct compilers for a wide variety of source languages and target machines using the same basic techniques.

12

SOFTWARE TECHNIQUES

Importance of SDLC: The SDLC helps you to define the "what", workout the "how" and develop and deliver a quality product by the "when". System development as defined in the SDLC version consists of seven key phases.

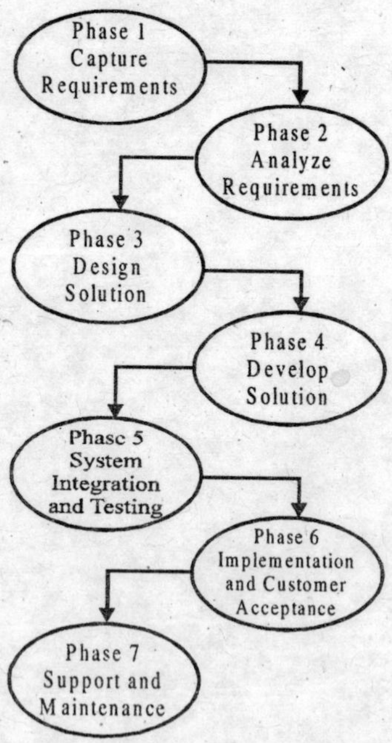

Diagrammatic Representation of the SDLC Phases

Phase 1: Capturing the Requirements: Involves sitting down with the client and doing a lot of listening, followed by a lot of writing. The end result is a document that defines what the customer requires in terms that the customer can understand.

Phase 2: Analyzing the Requirements: Involves developing a deeper understanding of the requirements and gather as much data about them as we can, to help us to design a solution.

Phase 3: Designing a Solution: This phase defines how the product will be built.

Phase 4: Developing a Solution: From the what and the how, the individual self-contained modules that make up the solution, is produced and tested.

Phase 5: System Integration and Testing: All the modules developed in phase 4 are taken and put them together as a complete solution. We confirm that we have a solid product that meets all the original requirements.

Phase 6: Implementation and Customer Acceptance: The solution is implemented and proved to meet the original requirements.

Phase 7: Support and Maintenance: After the customer accepts the software, we support and maintain it. This is a final phase of the life cycle.

Role of Waterfall Model

This model has five phases: The developer must complete each phase before the next phase begins. This model is named "Waterfall Model", because its diagrammatic representation resembles a cascade of waterfalls. The various phases are:

Requirement Analysis and Specification Phase: This phase understands the exact requirements of the customer and document them properly. The requirements describe the "what" of a system not the "how". This phase produces a large document, written in a natural language, contains a description of what the system will do without describing how it will be done. The resultant software requirement specification (SRS) document acts as contract between the developer and customer.

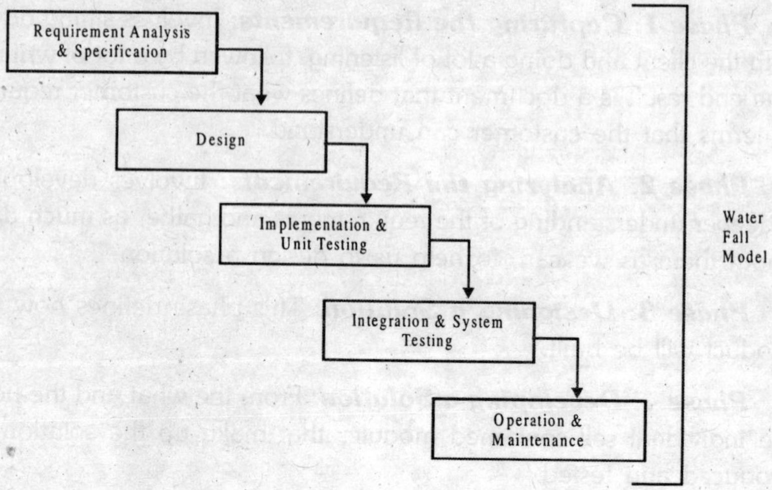

Design Phase: The goal of this phase is to transform the requirement specification into a structure that is suitable for implementation in some programming language. This work known as software design description (SDD) document must have information sufficient to begin the coding phase.

Implementation and Unit Testing Phase: During this phase, design is implemented. If the SDD is complete, the implementation or coding phase proceeds smoothly, because all the information needed by the software engineers is contained in the SDD.

Integration and System Testing Phase: This phase is to determine that each independent module is correctly implemented, which gives chance to determine that the interface between modules is also correct, and for this reason integration testing is performed. System testing involves the testing of the entire system. This is essential to build confidence in the developers before software is delivered to the customer or released in the market.

Operation and Maintenance Phase: Software maintenance is a task for development group, when the software is delivered to the customer's site, installed and is operational.

The purpose of this phase is to preserve the value of the software over time. This phase may span for 5 to 50 years whereas development may be 1 to 3 years.

This model reinforces the notion of "define before design" and "design before code". Working software is not available until relatively late in the process thus delaying the discovery of serious errors.

Importance of Spiral Model

Barry Boehm incorporated a "project risk" factor into a life cycle model.

The radial dimension of the model represents the cumulative costs. Each path around the spiral is indicative of increased costs. The angular dimension represents the progress made in completing each cycle. Each loop of the spiral from X-axis clockwise through 360° represents one phase. One phase is split roughly into four sectors of major activities.

1. *Planning:* Determination of objectives, alternatives and constraints.
2. *Risk Analysis:* Analyze alternatives and attempts to identify and resolve the risks involved.
3. *Development:* Product development and testing product.
4. *Assessment:* Customer evaluation.

During the first phase, planning is performed, risks are analyzed, prototypes are built and customers evaluate the prototype. During the second phase, a more refined prototype is built, requirements are documented and validated and customers are involved in assessing the new prototype. By the time third phase begins, risks are known and a somewhat more traditional development approach is taken.

An important feature of the spiral model is that each phase is completed with a review by the people concerned with the project (designers and programmers). This review consists of a review of all the products developed up to that point and includes the plans for the next cycle.

A. The advantage of this model is that it also incorporates software quality objectives into software development. The risk analysis and validation steps eliminate errors in the early phases of development.

B. This difficulties include lack of explicit process guidance in determining objectives, constraints, alternative: relying on risk assessment expertise; and provides more flexibility than required for many applications.

Metrics of Software

Definition: Software metrics can be defined as "The continuous application of measurement based techniques to the software development process and its products to supply meaningful and timely management information together with the use of those techniques to improve that process and its products".

Areas of Application

1. The most established area of software metrics is cost and size estimation techniques.
2. Controlling software development projects through measurement: This has become much more relevant with the increase in fixed price contracts and the use of penalty clauses by customers who deal with software developers.
3. The prediction of quality levels for software in terms of reliability.

Software Design: Process of applying various techniques and principles for purpose of defining a device, a process, a system insufficient detail to permit its physical realization.

It is a multi-step process that focuses on four distinct attributes of a programme.

1. Data structure.
2. Software architecture.
3. Interface representation.
4. Procedural detail.

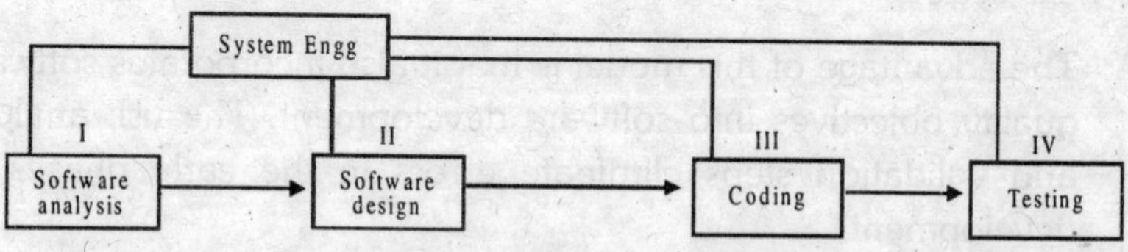

* Software design translates requirements into a representation of software that can be assessed for quality before code generation begins.

* Main goal of software design is to produce a model or representation of an entity that will be later built.
* Software design sits at the technical kernel of software engineering process and is applied regardless of software process model that is used.

Software Design Consists of

Data Design: Transforms information domain model created during analysis into data structures that will be required to implement; software.

Architectural Design: Defines relationship among major structural elements of programme.

Interface Design: Describes how software communicates with itself, with other systems and with humans who use it.

Procedural Design: Transforms structural components into procedural description of software components.

Role of Software Design

1. Design, code and test absorb 75% of software engineering.
2. It is software design which increases quality of software.
3. Design provides us with a representation of software which can be assessed for quality.
4. Design is the only way to translate customer requirements to finished software.
5. It serves as a foundation for all software engineering steps.
6. Without design we risk in building an unstable systems which can fail on small changes, a system difficult to test, difficult to check for quality.

Method of Software Design

1. Software design is an iterative process through which requirements are translated into a blueprint for constructing software.
2. Initially we have a holistic view of software, but after subsequent refinement we get a design that is very close to source code.

3. From project management point of view, software design is conducted in two steps:
   a Preliminary design: transformation of requirements into data and software architecture.
   b Detail design: refinements in architectural representations that lead to detailed data structure and algorithmic representation for software.

To evaluate the design representation, a design should

1. Exhibit a hierarchical organization that makes intelligent use of control among components of software.
2. Be modular; that is the software should be logically partitioned into components that perform specific functions and sub-functions.
3. Contain distinct and separable representation of data and procedure.
4. Lead to modules (*e.g.*, sub-routines or procedures) that exhibit independent functional characteristics.
5. Lead to interfaces that reduce the complexity of connections between modules and with the external environment.
6. Be derived using a repeatable method that is driven by information obtained during software requirements analysis.

Theories of Software Design

1. What criteria can be used to partition software into individual components?
2. How is function or data structure details separated from a conceptual representation of software?
3. Are there uniform criteria that define the technical quality of a software design?

Note: These concepts provide software designer with a foundation from which more sophisticated methods can be applied. Fundamental software design concepts provide necessary framework for getting a programme right.

Various fundamental software design concepts are:

1. Abstraction
2. Refinement
3. Modularity
4. Software architecture
5. Control hierarchy
6. Structural partitioning
7. Data structure
8. Software procedure
9. Information hiding

Designing the System

It is concerned with how system functionality to be provided by different components of a system. Activities involved in system design are:

Partition Requirements: There are several partitioning options and a number of alternatives may be produced in this stage.

Identify Sub-systems: Identifying sub-systems is carried out alongwith requirements partitioning. This process is also influenced by organizational or environment factors. It is also important to identify hardware to be built.

Assign Requirements to Sub-systems: Theoretically assigning requirements to sub-systems is a straight forward process if requirement partitioning is used to subsystem identification.

But practically there is never a clean match between above.

Specify Subsystem Functionality: This phase is a part of system design or sub-system development, or part of requirement specification for it.

There is no clear division between System design and Sub-system development.

Define Sub-system Interfaces: The interfaces that are provided and expected by each sub-system are defined in this phase. Once it is done, parallel development of sub-system becomes possible.

System design involves great deal of feedback and iteration between stages in this process. System design process also involves deciding which system capabilities are to be implemented in software and which in hardware.

Designing the Micro Level: It describes entities in design and their interfaces. These descriptions may be recorded in a data dictionary. It also describes control structure of design using PDL.

It produces a short design specification (Minispec) of each function, describes the function, its output and inputs.

The next stage of functional design process is to produce detailed design for each part of the design.

These detailed design should include control information and information about data structures manipulated.

Practical Designing

The design process for software systems often has two levels. At the first level the focus is on deciding which modules are needed for the system, the specification of these modules, and how the modules should be interconnected.

In Function Oriented design process the design is decomposed into a set of interacting units where each unit has a clearly defined function.

One of the best-known advocates of this method is Niklaus Wirth, the creator of PASCAL and a number of other languages. His special variety is called stepwise refinement, and it is a top down design method. This method works fine for small programmes.

Different kinds of objects that are to be manipulated by the Scheme interpreter (integer, real, *etc.*) need to be known by every module. Thus, if we need to add a type, every module needs to be altered. Needless to say, it should be avoided.

The refinement of each module continues until we reach the statement level of our programming language. At that point, we can describe the structure of our programme as a tree of refinements as in design top-down structure.

If a programme is created top-down, the modules become very specialized. In this design structure, each module is used by at most one.

ENCRYPTION AND CHECKING

Software Testing: It is a critical element of a software quality assurance and represents the ultimate review of specification, design and coding.

Software Testing Fundamentals: Software testing fundamentals define the overriding objectives of software testing.

In software testing, engineer creates a set of test case that is intended to demolish the software being built.

Objectives of Testing: The purpose of a good test case is to find an undiscovered error.

Testing Principles: A software engineer must understand basic principles that guide software testing before designing test cases.

Various testing principles given by Davis are:

1. All tests should be traceable to customer requirements.
2. Tests should be planned long before testing begins.
3. The Pareto principle applies to software testing.
4. Testing should begin "in the small" and progress towards testing "in the large".
5. Exhaustive testing is not possible.
6. To be most effective, testing should be conducted by third party.

Types of Testing: We have two types of testing.

TESTING VIA WHITE BOX

It is also called glass box testing. It is a test case design method that uses control structure of procedural design to derive test cases.

Using white box testing methods, we can derive test cases that:

* guarantee that all independent paths in a module have been exercised at least once.

* exercise logical decisions on their true and false sides.
* execute all loops at their boundaries and within their operational bounds.
* exercise internal data structures to assure their validity.

White box testing is needed because

* Logic errors and incorrect assumptions are inversely proportional to probability that a programme path will be executed.
* Typographical errors are random.

Black box testing may miss above errors, so we require white box testing.

Basis path testing is a white box testing technique. It enables test case designer to derive a logical complexity measure of procedural design. It can be used as a guide for defining basis set of execution path.

STANDARD OF SOFTWARE

Phillp Crosby, in his landmark book on quality, writes

"The problem of quality management is not what people don't know about it. The problem is what they think they do know ....".

Software quality assurance (SQÀ) is an umbrella activity that is applied throughout the software process.

Software Quality Assurance includes: (1) a quality management approach, (2) effective software engineering technology (methods and tools), (3) formal technical reviews that are applied throughout the software process, (4) a multitiered testing strategy, (5) control of software documentation and the changes made to it, (6) a procedure to ensure compliance with software development standards (when applicable) and (7) measurement and reporting mechanisms.

13

THE MULTIMEDIA

THE MEANING

Multimedia can be viewed as a form of communication when more than one media is utilized in an interactive manner.

The media elements include

(i) Text

(ii) Images

(iii) Video

(iv) Sound

THE ADVANTAGES

The benefits of multimedia are as follows

- Digital
- Interactive
- Modification
- Distribution

Benefits of Being Digital: All media elements are digitized separately and are stored on a single platform. There are softwares available which can be used to modify the media elements and process them. For example a sound file containing a narration can be mixed with background music. These digitized media elements are integrated through software called as authoring tool.

Role of Interactiveness

The interaction through the application can be such that people with varied interests can focus on their desired areas and thereby concentrate on topics of their choice. The viewer can get involved in the complete process of interaction while running designing multimedia application. These may be achieved through a response and feedback mechanism using interactive methods (such as cheek bones, radio buttons, *etc.*) or by using simple navigation prompts.

Uses of Modification

It is quite simple to modify and update the given information, as the information is stored on the computer in a digital format. There are plenty of softwares available which can be used for processing sound, video clips, still photographs, and text.

Advantages of Distribution: The ease in distribution of multimedia applications comes from the fact that all the media elements are stored on the same storage device. Multimedia applications are often delivered on a CD ROM, which has maximum capability to handle interactive data generated in any format. This clearly indicates that huge amount of data may be distributed through a fairly economical media.

Most multimedia software development coming these days are Internet ready. These allow the multimedia packages to be ported and distributed through the internet.

Need for Hardware

Minimum requirements for multimedia computer system are:

- PC-AT (80386)
- WINDOWS 3.1
- 16 MB RAM
- Hard Disk 1 GB
- 1 Floppy drive (1.44 MB)
- Mouse
- SVGA colour Monitor (14")

- Sound Card
- Video Card
- Speakers
- CD-ROM Drive

Multimedia elements and softwares requires enough disk space. Therefore more storage in terms of computers memory (Hard Disk and RAM) is desired.

A multimedia computer with a high configuration should include:

- Pentium III MHz with built in match co-processor
- 128 MB RAM
- 20 GB hard disk
- Multimedia Kit with sound card, speakers, microphones
- 128 bit Accelerated Graphics Card with 8 MB VRAM
- Mouse and Keyboard
- Windows 98/NT
- SVGA colour Monitor (15" or 17")

Need for Software

Software Platform: For IBM Pc's Windows environment (Win 3.1/95/98/NT) is essential as most of the multimedia application development software are Windows based.

Authoring Tool: This is software which allows integration of the different media elements—along with interactivity to create a final multimedia application software.

Example

1. Authorware 5 Attain
2. Director 7
3. Asymmetrix Toolbook

Still Image Capture: Still image can be captured using the software which comes along with a scanner.

Sound Processing: This software allows recording, editing, mixing and adding special effects for sound files.

Examples

1. Sound forge XP 4.0

Video Processing: This software allows capture, editing, mixing and special effects for video files.

Example

1. Adobe Premiere 5
2. Digital Video Producer

FIELDS OF APPLICATION

Multimedia can be put to varied uses from education to entertainment. Some of the areas on which Multimedia is used widely are as follows:

ROLE OF PUBLICITY

* TV commercials
* Multimedia advertising (Internet)
* Marketing Web page development

PROFITS OF ENTERTAINMENTS

* Film special effects
* Video animation and special effects
* Television
* Internet games development

BIBLIOGRAPHY

Apter, Michel: The New Technology of Education, *MacMillan, London, 1968.*

Awasthy, G.C.: Broadcasting in India, *Allied Publishers Private Ltd., Bombay, 1965*

Beiggs L.S.: Instruction Media, *Fitteburgh American Institute for Research, New York, 1966.*

Bloom, B.S.: Human Characteristics and School Learning, *Harper, New York, 1976.*

__________ : Taxonomy of Educational Objectives: The Classification of Educational Goals, *Mackey, New York, 1965.*

Brown, J.W. Lewis, R.B. & Hercleread. A.V.: Instruction: Materials and Methods, *McGraw Hill Book Co., New York, 1964.*

Chatterjee, R.K.: Mass Communication, *National Book Trust, New Delhi, 1999.*

Chauhan, S.S.: Textbook of Programmed Instruction, *Sterling Publishers, New Delhi, 1995.*

Cream, David: Explaining Teaching Machines and Programmes, *Featon, San Fracisco, 1961.*

Dale, Edgare: Audio-Visual Method in Teaching, *Dryden Rinehart and Winston, New York, 1969.*

Das, R.C.: Educational Technology, *Sterling Publishers, New Delhi, 1999*

Gagne, R.M.: The Conditions of Learning, *Holt Binsort and Winston, New York, 1965.*

Goodman, Loster: Educational Radio, *NCERT, New Delhi, 1977.*

Hancock, Alan: Planning for Educational Mass Media, *Longmans London, 1977.*

Kaul, B.N.: Studies in Distance Education, *Association of Indian Universities, IGNOU, New Delhi, 1988.*

Kinder, James. S.: Audio-Visual Materials and Techniques, *American Book Company, New York, 1959.*

Kothari, O.S.: Report of the Indian Education Commission, 1964-66, *Educational Pub., New Delhi, 1966.*

Kulkarni, S.S.: Introduction to Educational Technology, *Oxford Publishing Company, New Delhi, 1998.*

——————— : Teaching Learning Process: A Systems Analysis, *UNESCO, Bangkok, 1975.*

Lowman, J.: Mastering Techniques of Teaching, *Prentice Hall India, New Delhi, 1987.*

Matheson, Alastair: The Communication Materials and Methods, *McGraw Hill Book Company, New Delhi, 1964.*

Mc Luhan: Understanding Media, *Routludge, London, 1964.*

Mc. Whinnie, Donald: The Art of Radio, *Faber & Faber, London, 1961.*

Menon, Narayan: The Communication Revolution, *National Book Trust, New Delhi, 1976.*

Mohanty, J.: (ed) Studies in Distance Education, *Deep & Deep Publications, New Delhi, 1995.*

——————— : Dynamics of Higher Education in India, *Deep & Deep Publications, New Delhi, 1995.*

——————— : Educational Broadcasting: Radio & Television in Education, *Sterling Publishers, New Delhi, 1984.*

——————— : Educational Technology, *Deep & Deep Publications, New Delhi, 1991.*

——————— : Indian Education in the Emerging Society, *Sterling Publishers, New Delhi, 1982.*

——————— : The Open University: An Achievement in Educational Technology, *Oxford, New York, 2000.*

Mukhopadhyay, M.: Educational Technology Year Book, 1988, *AIAET, New Delhi, 1989.*

Munn, Norman L.: The Fundamentals of Human Adjustment, *Houghter Miffin Co., Boston, 1965.*

Ottaway A.K.: Education and Society, *Routldge and Kegan Paul, London, 1962.*

Rahman, S.A.: Study in Educational Television, *Concept Pub., Orissa, 1972.*

Rewntreu, Desk. Educational Technology in Curriculum Development, *Harper and Row Publishers, London, 1998.*

Sampath, K.: Introduction to Educational Technology, *Sterling, New Delhi, 1981.*

Schramm, Wilbur: Programmes Instruction: Today and Tomorrow, *The Funds of Advancement of Education, New York, 1962.*

————— : The Process and Effects of Mass Communication, *Uni of Illinois, Urbane, 1999.*

Sharma, G.D. & Ahmed, S.R.: Methodologies of Teaching in Colleges, *UNESCO UGC & NIEPA, New Delhi, 2000.*

Vedanayagam, G.: Teaching Technology for College Teachers, *Sterling Publishers, New Delhi, 1989.*